Distributions, Fourier Transforms
and
Some of Their Applications to Physics

World Scientific Lecture Notes in Physics Vol. 37

Distributions, Fourier Transforms
and
Some of Their Applications to Physics

Thomas Schücker

Institute for Theoretical Physics
University of Heidelberg

Wb World Scientific
Singapore • New Jersey • London • Hong Kong

Published by

World Scientific Publishing Co. Pte. Ltd.
P O Box 128, Farrer Road, Singapore 9128
USA office: 687 Hartwell Street, Teaneck, NJ 07666
UK office: 73 Lynton Mead, Totteridge, London N20 8DH

Library of Congress Cataloging-in-Publication data is available.

**DISTRIBUTIONS, FOURIER TRANSFORMS AND
SOME OF THEIR APPLICATIONS TO PHYSICS**

ISBN 981-02-0535-X

Printed in Singapore by JBW Printers & Binders Pte. Ltd.

to my brother John

Wer was zu sagen hat,
hat keine Eile.
Er läßt sich Zeit und sagt's
in einer Zeile.

Erich Kästner

If something's worth saying,
take your time.
And say it then,
in a single line.

translated by Mike Shiels

Preface

This book is based on lectures I held in Berkeley and Heidelberg during 1988-1990. It is intended for undergraduate students in mathematics, physics, and engineering. The only prerequisites on the mathematical side are linear algebra and real analysis while introductory courses on electrodynamics and quantum mechanics are needed to appreciate the physical motivations.

Distributions are often cited to examplify the interplay between mathematics and physics. Born from physical intuition, Huygens' principle in wave mechanics, point-like charges in electrodynamics and the wave function of a particle with well-defined position in quantum theory, distributions are today inherent to our way of thinking about physics. The manipulations involved when dealing with distributions were made rigorous by mathematicians and then distributions evolved quite independently to a tool used in such diverse branches of mathematics as the theory of linear differential equations, probability theory, group theory, and manifolds. All these branches offer results relevant to modern physics.

In the same way Fourier transforms constitute a solid link between physics and mathematics. Their applications to physics are almost universal: spectral analysis in acoustics, optics and electronics, "rhythm is the secret". The superposition principle, the particle-wave duality and the uncertainty principle are all encompassed by the Fourier transformation. Mathematicians use the Fourier transformation to solve linear differential equations with constant coefficients and have found fruitful generalizations in functional analysis.

The purpose of this text is to define distributions and Fourier transformations in $\mathbb{R}^n$ and to illustrate some of their applications to physics. Chapter one collects a few tools from complex analysis needed later to evaluate some real integrals. It is otherwise disconnected from the main stream.

In chapter two we motivate distributions with several physical examples, give their definition and discuss some of their properties. We prefer Temple's approach as introduced in Lighthill's book (Lighthill, 1962) to the original functional approach due to Laurent Schwartz. For our purpose Temple's definition of a distribution in terms of sequences of functions has two advantages: It uses Riemann's integral and by-passes most of the functional analysis as Lebesque's integral and continuous linear forms. It emphasizes the close relation between distributions and functions, facilitating the justification of some heuristic manipulations, e.g.

$$\delta(x^2 - m^2) = \frac{1}{2m}[\delta(x + m) - \delta(x - m)], \quad m > 0.$$

The Green function, the outstanding application of distributions, is defined and calculated for our leitmotiv example, the harmonic oscillator.

Chapter three is devoted to Fourier series of periodic functions and periodic distributions. In chapter four, motivated by a special limit of the Fourier series, we define the Fourier transform of functions. Then we generalize the Fourier transform to distributions. In this framework the Fourier series in turn becomes a special case of the Fourier transformation. Finally, the use of Fourier transforms to calculate Green functions is illustrated. Chapter five contains a discussion of some differential equations playing an important role in physics as well as their Green functions.

In chapter six we review linear algebra with special emphasis on infinite dimensional spaces and Hilbert spaces and explain how differential operators and Fourier transforms fit naturally into this frame. This point of view becomes fundamental in quantum mechanics. It also prepares the presentation of chapter seven, dealing with some systems of special functions which appear in almost every branch of physics.

It is a pleasure to thank Meinulf Göckeler and Vaughan Jones for help and advice.

Heidelberg, 1990   Thomas Schücker

Contents

Distributions, Fourier Transforms
and
Some of Their Applications to Physics

1

Functions of a complex variable

Complex numbers have played a fundamental part in physics since the discovery of quantum mechanics, where the wave function has real arguments but complex values. The theory of complex functions, i.e. functions of complex arguments, on the other hand, has found only subordinate applications of technical character in physics so far.

For convenience this chapter reviews briefly complex numbers and a small portion of complex analysis, essentially Green's real theorem in Cauchy's complex formulation. Its main purpose is the use of the residue theorem to evaluate definite integrals in one real variable which we shall need later on.

1.1 Complex numbers

Although sufficient for any numerical calculation, the field of rational numbers is geometrically incomplete. This incompleteness is cured by the extension to the real numbers $\mathbb{R}$. In the field $\mathbb{R}$ every Cauchy sequence converges, "the real axis has no holes". However, the reals are still algebraically incomplete, there are real polynomials that do not have real roots, e.g. $p(x) = x^2 + 1$. Therefore it is convenient to extend the real numbers to the field of complex numbers $\mathbb{C}$. A complex number z is a pair of real numbers x and y

$$z = (x, y). \tag{1.1}$$

Addition of complex numbers is defined componentwise

$$(x_1, y_1) + (x_2, y_2) := (x_1 + x_2, y_1 + y_2) \tag{1.2}$$

and is denoted by the same symbol $+$ as the addition of real numbers. The neutral element of addition, denoted by 0 as for real numbers, is

$$0 = (0, 0) \tag{1.3}$$

and the negative of the complex number (x, y) is

$$-(x, y) = (-x; -y). \tag{1.4}$$

1

So far $\mathbb{C}$ is just the two-dimensional vector space $\mathbb{R}^2$. Let us denote by

$$1 := (1, 0), \quad i := (0, 1) \tag{1.5}$$

its canonical basis. The product of two complex numbers, denoted by juxtaposition, is defined to be

$$(x_1, y_1)(x_2, y_2) := (x_1 x_2 - y_1 y_2, x_1 y_2 + x_2 y_1) \tag{1.6}$$

and as immediate consequences of this defintion we have
a) $1 = (1, 0)$ is the neutral element of multiplication:

$$1z = z1 = z, \tag{1.7}$$

b) the inverse of the complex number (x, y) is

$$(x, y)^{-1} =: \frac{1}{(x, y)} = \left(\frac{x}{x^2 + y^2}, \frac{-y}{x^2 + y^2} \right), \tag{1.8}$$

if $x^2 + y^2 \neq 0$.
c) commutativity of multiplication:

$$z_1 z_2 = z_2 z_1 \tag{1.9}$$

d) associativity of multiplication:

$$(z_1 z_2)z_3 = z_1(z_2 z_3), \tag{1.10}$$

e) and distributivity

$$z(z_1 + z_2) = zz_1 + zz_2. \tag{1.11}$$

Furthermore we have the important equation

$$i^2 = -1. \tag{1.12}$$

This means that i is a root of the complex polynomial $p(z) = z^2 + 1$, which, over the reals, has no root. In fact a fundamental theorem of complex numbers states that $\mathbb{C}$ is algebraically complete, i.e. every complex polynomial has complex roots.

During the extension from the real to the complex numbers we gained a fundamental property, algebraic completeness, but we also lost one: complex numbers are not ordered. Any set admits an ordering, indeed many and therefore arbitrary ones, e.g. the complex numbers may be ordered "alphabetically". However, in the context of fields we are interested in an ordering that respects the field structure, addition and multiplication. An ordering of a field is defined by a subset of "positive" numbers such that sums and products of positive numbers are positive. However, there cannot be a subset of positive complex numbers. $i^3 = -i$ would imply that i

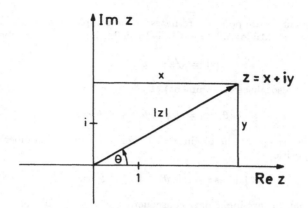

Fig. 1.1.   The complex plane

is both positive and negative. Inequalities make sense only for real, not for complex numbers.

A complex number $z = (x, y)$ can be decomposed with respect to the canonical basis 1 and i, $z = x1 + yi$. Persisting in our abuse of notation, denoting addition, multiplication and their neutral elements for complex numbers by the same symbol as for real numbers, we also drop the 1 in the above equation:

$$z = x1 + yi = x + iy, \tag{1.13}$$

thereby identifying the real number x with the complex number $(x, 0)$. The complicated multiplication law (1.6) can now be reconstructed from $i^2 = -1$ and the distributivity.

The complex number $z = (x, y) = x + iy$ can be represented by a point in a two-dimensional real plane, "the complex plane". The Cartesian coordinates of z are called real and imaginary part of z

$$\mathrm{Re}z := x, \quad \mathrm{Im}z := y. \tag{1.14}$$

A complex number is called real if its imaginary part vanishes and purely imaginary if its real part vanishes. For any complex number $z = x + iy$ we define its complex conjugate by

$$\bar{z} := x - iy. \tag{1.15}$$

In the complex plane complex conjugation is the reflection with respect to the real axis. It is also a field automorphism:

$$\overline{z_1 + z_2} = \overline{z_1} + \overline{z_2}, \tag{1.16}$$

$$\overline{z_1 z_2} = \overline{z_1}\,\overline{z_2}. \tag{1.17}$$

It is often convenient to use polar coordinates in the complex plane: The absolute value (or norm or magnitude) of $z = x + iy$ is by definition the real number

$$|z| := \sqrt{x^2 + y^2}. \tag{1.18}$$

The polar angle θ (also phase or argument) of z is given by

$$x =: |z| \cos\theta, \quad y =: |z| \sin\theta, \quad 0 \le \theta < 2\pi. \tag{1.19}$$

and is well defined except at the origin $z = 0$. For any complex number we have the polar decomposition

$$z = |z|(\cos\theta + i\sin\theta), \quad 0 \le \theta < 2\pi. \tag{1.20}$$

The following equations are immediate consequences of these definitions:

$$\text{Re}\, z = \frac{1}{2}(z + \bar{z}), \tag{1.21}$$

$$\text{Im}\, z = \frac{1}{2i}(z - \bar{z}), \tag{1.22}$$

$$|z|^2 = z\bar{z}. \tag{1.23}$$

Exercises

1. Show that the multiplication of complex numbers is associative and distributive.
2. Verify
$$\frac{a + ib}{c + id} = \frac{ac + bd}{c^2 + d^2} + i\frac{bc - ad}{c^2 + d^2}.$$
3. Show that the complex conjugation preserves addition and multiplication, equations (1.16) and (1.17).
4. Verify $|z|^2 = z\bar{z}$.

1.2 Roots

While the addition of complex numbers has a simple form in Cartesian coordinates, the multiplication takes a convenient form in polar coordinates:

$$
\begin{aligned}
z_1 z_2 &= |z_1|(\cos\theta_1 + i\sin\theta_1)|z_2|(\cos\theta_2 + i\sin\theta_2) \\
&= |z_1||z_2|[(\cos\theta_1\cos\theta_2 - \sin\theta_1\sin\theta_2) + i(\sin\theta_1\cos\theta_2 + \cos\theta_1\sin\theta_2)] \quad (1.24) \\
&= |z_1||z_2|[\cos(\theta_1 + \theta_2) + i\sin(\theta_1 + \theta_2)].
\end{aligned}
$$

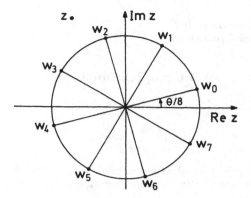

Fig. 1.2   The 8th roots of z

"Under multiplication absolute values multiply and polar angles are added modulo 2π." In particular

$$|z_1 z_2| = |z_1||z_2|. \tag{1.25}$$

On the other hand note the triangle inequality

$$|z_1 + z_2| \le |z_1| + |z_2|. \tag{1.26}$$

Definition: An n-th root w of $z \in \mathbb{C}, n = 1, 2, 3..$ is a complex number w such that

$$w^n = z. \tag{1.27}$$

Iterating (1.25) we get

$$(\cos\theta + \mathrm{i}\sin\theta)^n = \cos n\theta + \mathrm{i}\sin n\theta, \quad n \in \mathbb{N} \tag{1.28}$$

and if z is written in polar coordinates (1.20) an obvious n-th root of z, the so-called principal root is given by

$$w_0 = \sqrt[n]{|z|}(\cos\frac{\theta}{n} + \mathrm{i}\,\sin\frac{\theta}{n}). \tag{1.29}$$

More generally, if z is non-zero, there are n distinct n-th roots of z:

$$w_k = \sqrt[n]{|z|}\left(\cos\frac{\theta + 2\pi k}{n} + \mathrm{i}\,\sin\frac{\theta + 2\pi k}{n}\right), k = 0, 1, 2, \cdots n - 1. \tag{1.30}$$

For example $z = -1$, its absolute value is 1, its polar angle $\theta = \pi$ and its second roots are $w_0 = \mathrm{i}$, $w_1 = -\mathrm{i}$

Exercises

1. Prove the triangle inequality.
2. Calculate all third roots of -1.

1.3 Complex functions

A complex function $f(z)$ is a map from the complex numbers into the complex numbers

$$f : \mathbb{C} \longrightarrow \mathbb{C}$$
$$z \mapsto f(z).$$

Alternatively, f may be viewed as a function from $\mathbb{R}^2$ into $\mathbb{R}^2$ by decomposing argument and value into real and imaginary part, $z = x + iy$ and

$$f(z) = u(x, y) + iv(x, y). \tag{1.31}$$

Examples are the complex conjugation

$$f(z) = \bar{z}, \quad u(x, y) = x, \quad v(x, y) = -y, \tag{1.32}$$

polynomials, e.g.

$$f(z) = z^2 + 1, \quad u(x, y) = x^2 - y^2 + 1, \quad v(x, y) = 2xy \tag{1.33}$$

or the n-th principal root. Note that the first two examples are continuous functions while the principal root is discontinuous along the positive real axis. A most important complex function is the exponential. We extend the definition of the real exponential function to the complex plane by

$$e^{x+iy} := e^x(\cos y + i \sin y), \quad \text{"Euler's formula"}. \tag{1.34}$$

A remarkable property of this definition is that the complex exponential also satisfies the functional equation

$$e^0 = 1, \quad e^{z_1+z_2} = e^{z_1} e^{z_2} \tag{1.35}$$

i.e. it defines a group homomorphism from the additive to the multiplicative group of complex numbers. The last equation follows from Euler's formula using the trigonometric angle sum relations as in equation (1.24). Further consequences are:

$$|e^z| = e^{\mathbf{Re} z}, \tag{1.36}$$

in particular

$$|e^{ix}| = 1, \quad x \in \mathbb{R}, \tag{1.37}$$

$$\cos x = \frac{1}{2}(e^{\mathrm{i}x} + e^{-\mathrm{i}x}), \qquad (1.38)$$

$$\sin x = \frac{1}{2\mathrm{i}}(e^{\mathrm{i}x} - e^{-\mathrm{i}x}). \qquad (1.39)$$

One major advantage in using exponential functions with purely imaginary argument rather than trigonometric functions in physical problems concerned with periodic arrangements is that the simple functional equation (1.35) replaces many complicated trigonometric identities. For example:

$$\begin{aligned} \cos 2x = \mathrm{Re}e^{2\mathrm{i}x} &= \mathrm{Re}(e^{\mathrm{i}x}e^{\mathrm{i}x}) = \mathrm{Re}e^{\mathrm{i}x}\,\mathrm{Re}e^{\mathrm{i}x} - \mathrm{Im}e^{\mathrm{i}x}\,\mathrm{Im}e^{\mathrm{i}x} \\ &= \cos^2 x - \sin^2 x. \end{aligned} \qquad (1.40)$$

Exercise

1. Solve the damped harmonic oscillator

$$\frac{d^2}{dt^2}x(t) + 2\lambda\frac{d}{dt}x(t) + \omega_0^2 x(t) = 0,$$

λ and real ω_0, with initial conditions

$$x(0) = x_0, \qquad \frac{dx}{dt}(0) = 0$$

using the complex exponential.

1.4 Differentiation

We generalize the derivative from real to complex functions.

Definition: Let f be a complex function. If the limit

$$\lim_{\Delta z \to 0} \frac{f(z + \Delta z) - f(z)}{\Delta z}$$

exists for all sequences Δz approaching zero, it is called the (complex) derivative of f at z, denoted by $f'(z)$ or $\frac{df}{dz}(z)$.

The definition of the complex derivative is much more restrictive than the one of partial derivatives of the component functions $u(x, y)$ and $v(x, y)$, where horizontal limits, $\Delta z = \Delta x$ real, and vertical limits, $\Delta z = \mathrm{i}\Delta y$ purely imaginary, may well differ. Let us illustrate this point by a complex function, which is not differentiable

8

in the complex sense. Consider the complex conjugation $f(z) = \bar{z}$ at the origin $z = 0$. The limit to examine is

$$\lim_{\Delta z \to 0} \frac{f(0 + \Delta z) - f(0)}{\Delta z} = \lim_{\Delta z \to 0} \frac{\overline{\Delta z}}{\Delta z}. \tag{1.41}$$

This limit is 1 for Δz approaching 0 on the real axis, but -1 for Δz approaching 0 on the imaginary axis. Although its component functions $u(x, y) = x$ and $v(x, y) = -y$ have well–defined partial derivatives, the complex conjugation is not differentiable in the complex sense, because if we spiral down to the origin

$$\Delta z = \frac{1}{n} e^{in\pi/2}, \quad n = 1, 2, 3 \ldots \tag{1.42}$$

the limit (1.44) oscillates between -1 and +1 and fails to exist.

For any complex function $f = u + iv$ complex differentiability at $z = x + iy$ means that all simultaneous limits

$$f'(z) = \lim_{\substack{\Delta x \to 0 \\ \Delta y \to 0}} \frac{[u(x + \Delta x, y + \Delta y) - u(x, y)] + i[v(x + \Delta x, y + \Delta y) - v(x, y)]}{\Delta x + i\Delta y} \tag{1.43}$$

exist and are all equal. In particular for $\Delta y \equiv 0$

$$f' = \frac{\partial u}{\partial x} + i \frac{\partial v}{\partial x} \tag{1.44}$$

and for $\Delta x = 0$

$$f' = \frac{\partial v}{\partial y} - i \frac{\partial u}{\partial y}. \tag{1.45}$$

Equating real and imaginary part we arrive at the so–called Cauchy–Riemann equations

$$\frac{\partial u}{\partial x} = \frac{\partial v}{\partial y}, \tag{1.46}$$

$$\frac{\partial u}{\partial y} = -\frac{\partial v}{\partial x}. \tag{1.47}$$

They are necessary and also sufficient equations for a differentiable function f to admit a complex derivative.

In the above example $f(z) = \bar{z}$ the first Cauchy-Riemann equation is not satisfied in any point, the complex conjugation admits nowhere a complex derivative.

Definition: A complex function f is said to be holomorphic (or regular or analytic) at a point $z \in \mathbb{C}$ if it possesses a complex derivative in every point of some neighbourhood of z. The function is called holomorphic if it is holomorphic in every point of the complex plane.

A fundamental theorem of complex functions states that a holomorphic function admits complex derivatives of any order and that its Taylor series converges to this function everywhere, whence the name analytic. Both statements fail of course in the case of real functions.

On the other hand the following theorems of real analysis hold true in the complex setting:

a) If f and g are holomorphic functions then so are their sum, complex multiples, product and composition and

$$(f + g)' = f' + g', \tag{1.48}$$
$$(cf)' = cf', \quad c \in \mathbb{C}, \tag{1.49}$$
$$(fg)' = f'g + fg', \tag{1.50}$$

$$\frac{d}{dz}(g(f(z))) = g'(f(z))f'(z), \quad \text{"chain rule".}$$

b) For any positive integer n $\quad f(z) = z^n$ is holomorphic and

$$\frac{d}{dz}(z^n) = nz^{n-1}. \tag{1.51}$$

c) The exponential is holomorphic and

$$\frac{d}{dz}e^z = e^z. \tag{1.52}$$

The proofs of a) and b) are literally the same as for functions of one real variable. To prove c) we first verify the Cauchy-Riemann conditions and then use (1.44).

A final remark: by partial differentiating the Cauchy-Riemann equations we find that both component functions of a holomorphic function are harmonic:

$$\left(\frac{\partial^2}{\partial x^2} + \frac{\partial^2}{\partial y^2}\right)u = \left(\frac{\partial^2}{\partial x^2} + \frac{\partial^2}{\partial y^2}\right)v = 0. \tag{1.53}$$

Exercises

1. Verify the Cauchy-Riemann condition for $f(z) = z^3$.
2. Calculate the derivative of the exponential by equations (1.44) and (1.45).
3. Verify equation (1.53).
4. Is the function

$$f(z) := \begin{cases} e^{-1/z^4}, & z \neq 0 \\ 0, & z = 0 \end{cases}$$

holomorphic at the origin? Are the Cauchy-Riemann equations satisfied at the origin?

10

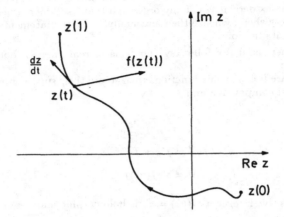

Fig. 1.3.   A curve in the complex plane

1.5 Integration

We would like now to define an integral of a complex function. A complex function can be considered as vector field in the complex plane. This point of view suggests to use some line integral. However, we replace the scalar product appearing in the ordinary line integral of a vector field by the product of complex numbers.

Definition: Let C be an (oriented, piecewise smooth) curve in the complex plane and $f = u + iv$ a complex function. The (complex) integral of f along C is

$$\int_C f := \int_C f(z)dz := \int_C (udx - vdy) + i \int_C (vdx + udy) \qquad (1.54)$$

if the real line integrals on the rhs exist.

This definition is easy to remember by putting $dz = dx + idy$ and by decomposing $f\ dz$ into real and imaginary part. Let us recall the definition of the line integral of a real function. Choose a parameter representation $z(t)$ of the curve C

$$\begin{aligned} z : [0,1] &\longrightarrow \mathbb{C} \\ t &\mapsto z(t) = x(t) + iy(t). \end{aligned} \qquad (1.55)$$

Then one of the integrals in (1.54) is

$$\int_C udx := \int_0^1 u(x(t), y(t)) \frac{dx}{dt} dt. \qquad (1.56)$$

Example: Let C be the upper half circle of radius R centered around the origin, oriented in a counter clockwise sense: We choose the parameter representation

$$z(t) = R\,e^{it}, \quad t \in [0, \pi] \tag{1.57}$$

or

$$x(t) = R\cos t, \quad y(t) = R\sin t. \tag{1.58}$$

Let $f(z) = z^2$ with $u = x^2 - y^2$ and $v = 2xy$. Then

$$
\begin{aligned}
\int_C f &= \int_C (x^2 - y^2)dx - 2\int_C xy\,dy + i\int_C 2xy\,dx + i\int_C (x^2 - y^2)dy \\
&= \int_0^\pi (R^2\cos^2 t - R^2\sin^2 t)(-R\sin t)dt - \int_0^\pi 2R^2\cos t\sin t R\cos t\;dt \\
&\quad + i\int_0^\pi 2R^2\cos t\sin t(-R\sin t)dt + i\int_0^\pi (R^2\cos^2 t - R^2\sin^2 t)R\cos t dt \\
&= -3R^3\int_0^\pi \cos^2 t\sin t dt + R^3\int_0^\pi \sin^3 t dt \\
&= -4R^3\int_0^\pi \cos^2 t\sin t dt + R^3\int_0^\pi \sin t dt \\
&= 4R^3\int_1^{-1} w^2 dw + R^3\int_0^\pi \sin t dt = -\frac{8}{3}R^3 + 2R^3 = -\frac{2}{3}R^3.
\end{aligned}
\tag{1.59}
$$

The same calculation becomes more transparent in the compact complex notation:

$$z(t) = Re^{it}, \; dz = iRe^{it}dt \quad \text{and} \quad f(z(t)) = R^2 e^{2it}.$$

In this form the integral is not plagued by trigonometric functions. (Compare the remark following equation (1.39).)

$$
\begin{aligned}
\int_C z^2 dz &= \int_0^\pi R^2 e^{2it} iR\,e^{it}dt = R^3\int_0^\pi e^{3it}idt \\
&= \frac{R^3}{3}e^{3it}\Big|_0^\pi = -\frac{2}{3}R^3.
\end{aligned}
\tag{1.60}
$$

From of the real line integral the complex integral inherits the following properties:

a) Let f and g be two complex functions, C a curve and k a complex number. Then

$$\int_C (f + g) = \int_C f + \int_C g, \tag{1.61}$$

$$\int_C kf = k\int_C f \tag{1.62}$$

12

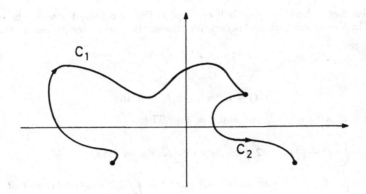

Fig. 1.4   The addition of curves

if all integrals exist.

b) If C_1 and C_2 are curves such that the end point of C_1 coincides with the initial point of C_2, then

$$\int_{C_1+C_2} f = \int_{C_1} f + \int_{C_2} f \qquad (1.63)$$

where $C_1 + C_2$ is explained in fig. 1.4.

c) If $-C$ denotes the curve having the same trace as C but opposite orientation, then

$$\int_{-C} f = -\int_C f. \qquad (1.64)$$

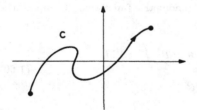   

Fig. 1.5.   The negative of a curve

d) If L is the length of the curve C

$$L := \int_0^1 \sqrt{\left(\frac{dx}{dt}\right)^2 + \left(\frac{dy}{dt}\right)^2}\, dt \qquad (1.65)$$

then

$$\left| \int_C f \right| \leq L \max_C |f(z)|. \tag{1.66}$$

Recall from mechanics that a force field f in $\mathbb{R}^3$ which is conservative

$$\text{curl } f = 0 \tag{1.67}$$

does not do work around closed trajectories, the line integral of f over any closed curve c vanishes due to Stokes' theorem

$$\oint_C f \cdot dx = \int\int_S \text{curl } f \cdot dS = 0 \tag{1.68}$$

where S is any surface bounded by C. In two dimensions the Cauchy-Riemann equations are the analogue of equation (1.67) with respect to the complex integral and Stokes' theorem in two dimensions is better known as

Green's theorem: Let $P(x, y)$ and $Q(x, y)$ be two real functions defined in a simply connected domain D of $\mathbb{R}^2$ (i.e. without holes). Suppose both functions have continuous first partial derivatives in D. Let C be a closed curve in D, oriented counter-clockwise and let S be the interior of C. Then

$$\oint_C (Pdx + Qdy) = \int\int_S \left(\frac{\partial Q}{\partial x} - \frac{\partial P}{\partial y} \right) dxdy. \tag{1.69}$$

In terms of complex functions Green's theorem yields

Cauchy's theorem: If $f(z)$ is a holomorphic function defined in a simply connected domain D of the complex plane and C a closed curve in D, then

$$\oint_C f(z)dz = 0. \tag{1.70}$$

Example: Let C be the circle of radius R centered at the origin and $f(z) = z^n$, $n = 1, 2, 3...$

$$\int_C z^n dz = R^{n+1} \int_0^{2\pi} e^{(n+1)rit} i dt = R^{n+1} \frac{1}{n+1} e^{(n+1)it} \Big|_0^{2\pi} = 0. \tag{1.71}$$

As important consequence of Cauchy's theorem we see that the integral of a holomorphic function does not depend on the curve C but only on the initial and final point of C.

Definition: Let f be a function holomorphic in a simply connected domain D and let a and b be complex numbers in D. The integral of f from a to b

$$\int_a^b f := \int_a^b f(z)dz \tag{1.72}$$

is the integral of f over any curve in D connecting a to b.

Now the complex integral resembles the real integral and we also have a

Fundamental theorem of differentiation and integration: Let $f(z)$ be holomorphic in a simply connected domain D and let z_0 be a point in D. Then the function

$$F(z) := \int_{z_0}^{z} f(\zeta)d\zeta \qquad (1.73)$$

is also holomorphic in D and

$$F' = f. \qquad (1.74)$$

Proof: Let $z = x + iy$, $z_0 = x_0 + iy_0$, $f = u + iv$, $F = U + iV$. Then

$$U = \int_{(x_0,y_0)}^{(x,y)} (udx - vdy) \qquad (1.75)$$

and

$$V = \int_{(x_0,y_0)}^{(x,y)} (vdx + udy). \qquad (1.76)$$

Using the fundamental theorem for real line integrals we verify the Cauchy-Riemann equations

$$\frac{\partial U}{\partial x} = u = \frac{\partial V}{\partial y} \quad \text{and} \quad \frac{\partial U}{\partial y} = -v = -\frac{\partial V}{\partial x}. \qquad (1.77)$$

Therefore F is holomorphic and $F' = \partial U/\partial x + i\partial V/\partial x = u + iv = f$.

In our example $\int_C z^2 dz$ over the upper half circle, the calculation can now be simplified by computing the integral from R to $-R$ along the real line and even better we can bypass the choice of a curve altogether and get the integral from a primitive just as in the real case.

$$\int_{R}^{-R} z^2 dz = \frac{z^3}{3}\Big|_{R}^{-R} = -\frac{2}{3}R^3. \qquad (1.78)$$

Exercises

1. Calculate $\int_C z^3 dz$ where C is the straight line from 1 to i without using the fundamental theorem.
2. Prove the inequality (1.66).
3. Prove Cauchy's theorem.
4. Compute $\int_1^i z^3 dz$.

1.6 Residues

We now consider examples where the domain of holomorphy has holes.

Definition: Let f be a complex function. A complex number z_0 is called an isolated singularity of f if there is a neighbourhood D of z_0 in the complex plane such that f is holomorphic at every point of D except at z_0 itself.

Example: The origin is an isolated singularity of $1/z^n$, $n = 1, 2, 3....$

Since $D - \{z_0\}$ is not simply connected we expect Cauchy's theorem to fail. Indeed if C is the circle of radius R around z_0 parametrized by

$$z(t) = z_0 + Re^{it}, \quad t \in [0, 2\pi] \tag{1.79}$$

then

$$\oint_C \frac{dz}{z - z_0} = \int_0^{2\pi} \frac{1}{Re^{it}} iRe^{it} dt = i \int_0^{2\pi} dt = 2\pi ri \tag{1.80}$$

The integral does not vanish, however, its value does not depend on the radius R. More generally, for any function f with isolated singularity z_0 any closed counterclockwise curve K surrounding z_0 once and lying in the domain D where f is holomorphic the integral of f over K does not depend on the choice of such a curve.

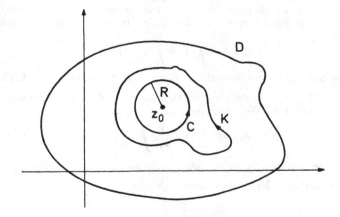

Fig. 1.6. Curves around an isolated singularity

16

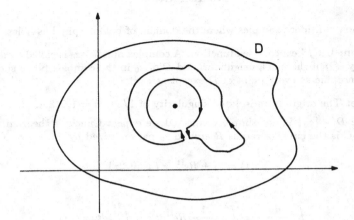

Fig. 1.7.   To apply Cauchy's theorem

The proof relies on a simple universal trick. Choose a circular curve C with small enough radius so that C lies inside K. Construct the curve $\tilde{K}$ by cutting a small channel between K and C as in fig. 1.7.

By construction $\tilde{K}$ is contained in a simply connected subset of D. By Cauchy's theorem $\oint_{\tilde{K}} f = 0$. In the limit where the channel width ϵ tends to zero, the two integrals along the channel cancel because they have opposite orientations and we remain with

$$0 = \lim_{\epsilon \to 0} \oint_{\tilde{K}} f = \oint_{K} f - \oint_{C} f \qquad (1.81)$$

where the minus sign comes from the clockwise orientation of the circle C.

Definition: Let $f(z)$ be a complex function holomorphic in a neighbourhood D of z_0 except possibly at z_0. We define the residue of f at z_0 by

$$\operatorname{Res} f(z)\big|_{z_0} := \frac{1}{2\pi i} \oint f(z)dz \qquad (1.82)$$

where the integral runs over any counter clockwise curve surrounding z_0 once.

The normalization has been chosen such that

$$\operatorname{Res} \frac{1}{z - z_0}\bigg|_{z_0} = 1. \qquad (1.83)$$

If f is holomorphic at z_0 then by Cauchy's theorem

$$\operatorname{Res} f(z)\bigg|_{z_0} = 0. \qquad (1.84)$$

On the other hand, an isolated singularity can have vanishing residue, e.g.

$$\text{Res} \left. \frac{1}{(z - z_0)^n} \right|_{z_0} = 0, \quad n = 2, 3, 4... \tag{1.85}$$

The main result of this section is the residue theorem. It is a two-dimensional analogue of Gauss' law in electrodynamics expressing the additivity of electric charge.

Residue theorem: If $f(z)$ is holomorphic on and inside a closed counter-clockwise curve C except for a finite number of isolated singularities $z_1, z_2, ...z_n$, all located inside C, then

$$\oint_C f = 2\pi i \sum_{k=1}^{n} \text{Res} \left. f \right|_{z_k}. \tag{1.86}$$

The proof uses the channel trick as illustrated in fig. 1.8.

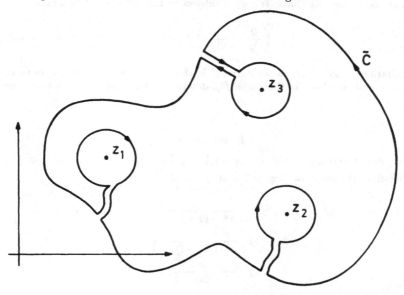

Fig. 1.8.   The residue theorem

In later applications the lhs of equation (1.86) will be expressed as a real integral we want to compute. There is a variety of methods to evaluate the residues appearing on the rhs without doing integrals. For our purpose the following theorem, which we cite without proof, will be sufficient.

Quotient theorem: Let f and g be two functions holomorphic in a neighbourhood of z_0 and satisfying

$$f(z_0) \neq 0, \quad g(z_0) = 0, \quad g'(z_0) \neq 0.$$

Then

$$\text{Res} \left. \frac{f}{g} \right|_{z_0} = \frac{f(z_0)}{g'(z_0)}. \tag{1.87}$$

An astonishing corollary is obtained by putting $g(z) = z - z_0$:

$$f(z_0) = \text{Res} \left. \frac{f}{z - z_0} \right|_{z_0} = \frac{1}{2\pi i} \oint_C \frac{f(z)}{z - z_0} dz. \tag{1.88}$$

It tells us that the values of a holomorphic function inside a given closed curve are already determined by its values on this curve.

This situation is analogous to the static soap film or drum skin suspended freely from a closed frame. Its shape $f(x, y)$ is subject to the differential equation

$$\left(\frac{\partial^2}{\partial x^2} + \frac{\partial^2}{\partial y^2} \right) f = 0 \tag{1.89}$$

as in equation (1.53). The closed curve C is the projection of the frame onto the $x - y$ plane and the shape of the frame $f|_C$ determines the shape of the whole drum skin.

Exercises

1. Calculate the residue of z^n at the origin for any integer n.
2. Calculate the residues at $z_0 = 1$ and $z_0 = i$ of

$$a) \quad \frac{1}{z^2 - (1 + i)z + i}$$

$$b) \quad \frac{1}{z^2 + (1 - i)z - i}$$

$$c) \quad \frac{1}{z^2 - 2iz - 1}.$$

1.7 Applications to the real world

In view of some definite integrals of functions with real argument which we shall encounter later, let us discuss a few applications of the residue theorem.

$$a) \quad \int_0^{2\pi} \frac{dt}{a + \sin t}, \quad a > 1$$

We rewrite this integral as a complex integral along the unit circle C of an appropriate function and put

$$z(t) = e^{it}, \tag{1.90}$$

$$dt = \frac{dz}{iz} \tag{1.91}$$

and from equation (1.39)

$$\sin t = \frac{1}{2i}\left(z - \frac{1}{z}\right). \tag{1.92}$$

Then our integral becomes

$$\int_0^{2\pi} \frac{dt}{a + \sin t} = \oint_C \frac{dz}{iz[a + \frac{1}{2i}(z - 1/z)]} = \frac{1}{i} \oint_C \frac{2i\,dz}{z^2 + 2iaz - 1}. \tag{1.93}$$

The integrand has only one isolated singularity inside the unit circle

$$z_0 = -ia + i\sqrt{a^2 - 1} \tag{1.94}$$

and by the residue theorem

$$\int_0^{2\pi} \frac{dt}{a + \sin t} = 2\pi \operatorname{Res} \frac{2i}{z^2 + 2iaz - 1}\Big|_{z_0}. \tag{1.95}$$

By the quotient theorem this residue is

$$\frac{i}{z_0 + ia} = \frac{1}{\sqrt{a^2 - 1}} \tag{1.96}$$

and our integral

$$\int_0^{2\pi} \frac{dt}{a + \sin t} = \frac{2\pi}{\sqrt{a^2 - 1}}. \tag{1.97}$$

This example generalizes to integrals of the form

$$\int_0^{2\pi} R(\sin\ t, \cos\ t)dt$$

where $R(x, y)$ is a rational function with no singularities on the unit circle.

$$\text{b)} \quad \int_{-\infty}^{+\infty} \frac{dx}{x^2 + a^2}, \quad a > 0$$

Although the integrand has an elementary primitive let us evaluate the integral by extending it to the complex plane. Let C be the closed curve along the real axis

20

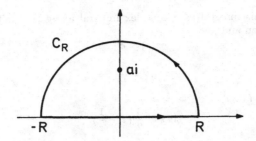

Fig. 1.9   Integrating along a curve

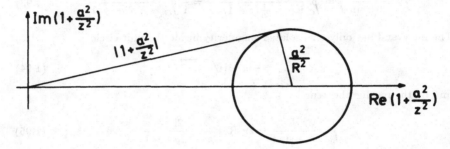

Fig. 1.10   An estimate

between $-R$ and R and around the upper half circle C_R of radius R as illustrated in fig. 1.9.

First we prove that

$$\int_{-\infty}^{+\infty} \frac{dx}{x^2 + a^2} = \lim_{R \to \infty} \int_C \frac{dz}{z^2 + a^2} = \lim_{R \to \infty} \int_{-R}^{R} \frac{dx}{x^2 + a^2} + \lim_{R \to \infty} \int_{C_R} \frac{dz}{z^2 + a^2}. \quad (1.98)$$

We have to show that the second limit vanishes: If z is on the half circle C_R with radius $R > \sqrt{2}a$, then $1 + a^2/z^2$ lies on the circle centered at 1 with radius $a^2/z^2 < \frac{1}{2}$ and by fig. 1.10:

$$|1 + \frac{a^2}{z^2}| > \frac{1}{2}. \quad (1.99)$$

Then

$$\left| \frac{1}{z^2 + a^2} \right| = \frac{1}{|z|^2} \frac{1}{|1 + a^2/z^2|} < \frac{2}{R^2} \quad (1.100)$$

and using inequality (1.66)

$$\left| \int_{C_R} \frac{dz}{z^2 + a^2} \right| \leq \pi R \max_{C_R} \left| \frac{1}{z^2 + a^2} \right| < \pi R \frac{2}{R^2} = \frac{2\pi}{R} \xrightarrow[R \to \infty]{} 0. \quad (1.101)$$

For $R > \sqrt{2}a$ the function $1/(z^2 + a^2)$ has one isolated singularity inside the curve C, at $z_0 = ia$ and

$$\int_{-\infty}^{+\infty} \frac{dx}{x^2 + a^2} = 2\pi i \operatorname{Res} \frac{1}{z^2 + a^2}\Big|_{ia} = 2\pi i \frac{1}{2z}\Big|_{ia} = \frac{\pi}{a}. \qquad (1.102)$$

This technique works for convergent integrals of the form

$$\int_{-\infty}^{+\infty} R(x)dx,$$

$R(x)$ being a rational function without singularities on the real axis.

$$c) \quad \int_{-\infty}^{+\infty} \frac{e^{ix}}{x}dx$$

Let us continue the integrand to the complex plane and integrate along the closed curve as in fig. 1.11:

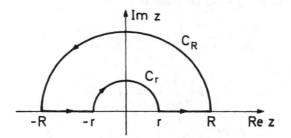

Fig. 1.11   Avoiding the pole

$$\int_{-R}^{-r} + \int_{r}^{R} + \int_{C_r} + \int_{C_R} \frac{e^{iz}}{z}dz = 0 \qquad (1.103)$$

by Cauchy's theorem.

First we show that

$$\lim_{R\to\infty} \int_{C_R} \frac{e^{iz}}{z}dz = 0. \qquad (1.104)$$

This time the length of C_R increases at the same rate as the integrand decreases with R and we have to work a little harder than in the preceding example. Integration by parts applies to holomorphic functions just as to real functions because it only relies on the product rule for differentiation and the fundamental theorem of differentiation and integration. In the case at hand integration by parts yields:

$$\int_{C_R} \frac{e^{iz}}{z} dz = \frac{e^{iz}}{iz}\Big|_{-R}^{R} + \int_{C_R} \frac{e^{iz}}{iz^2} dz$$
$$= \frac{2}{iR} \cos R + \int_{C_R} \frac{e^{iz}}{iz^2} dz. \tag{1.105}$$

Now as R tends to infinity also the second term tends to zero as in b) since $|e^{iz}| < 1$ in the upper half plane.

Next we show that

$$\lim_{r \to 0} \int_{C_r} \frac{e^{iz}}{z} dz = -\pi i. \tag{1.106}$$

The exponential e^{iz} is holomorphic at the origin and there it takes the value 1. Thus

$$|e^{iz} - 1| < \epsilon|z| \tag{1.107}$$

for all z with sufficiently small absolute values and

$$\left| \int_{C_r} \frac{e^{iz} - 1}{z} dz \right| \le \pi r \max_{C_r} \frac{|e^{iz} - 1|}{|z|} \le \pi r \epsilon. \tag{1.108}$$

Consequently

$$\lim_{r \to 0} \int_{C_r} \frac{e^{iz} - 1}{z} dz = 0 \tag{1.109}$$

and

$$\lim_{r \to 0} \int_{C_r} \frac{e^{iz}}{z} dz = \lim_{r \to 0} \int_{C_r} \frac{e^{iz} - 1}{z} dz + \lim_{r \to 0} \int_{C_r} \frac{dz}{z}$$
$$= \lim_{r \to 0} -i \int_0^\pi dt = -\pi i. \tag{1.110}$$

In the limit $R \to \infty$ and $r \to 0$ the first two terms in equation (1.103) tend to the integral we are after

$$\int_{-\infty}^{+\infty} \frac{e^{ix}}{x} dx = \pi i. \tag{1.111}$$

This technique can be generalized to other integrals of the form

$$\int_{-\infty}^{+\infty} e^{ix} f(x) dx$$

which will appear in Fourier transforms.

Exercises

1. Compute
$$\int_0^{2\pi} \frac{dt}{1 - 2c\cos t + c^2}, \quad c \in \mathbb{C}, \quad |c| \neq 1.$$

2. Compute
$$\int_{-\infty}^{+\infty} \frac{x^2}{x^6 + 1} dx.$$

3. Compute
$$\int_{-\infty}^{+\infty} \frac{e^{ikx}}{x^2 + a^2} dx, \quad a, k \in \mathbb{R}, \quad a \neq 0.$$

4. Calculate the Fresnel integrals
$$\int_0^\infty \sin(x^2)dx \quad \text{and} \quad \int_0^\infty \cos(x^2)dx.$$

Hint: Integrate e^{-z^2} over an eighth of a circle.

2

Distributions

2.1 Motivation

Consider a bouncing ball with mass M. Let $x(t)$ be its height over the floor from which it bounces off with velocity v_0

$$x(t) = \begin{cases} -\frac{1}{2}gt^2 - v_0 t, & -2v_0/g < t \leq 0 \\ -\frac{1}{2}gt^2 + v_0 t, & 0 \leq t < 2v_0/g. \end{cases} \qquad (2.1)$$

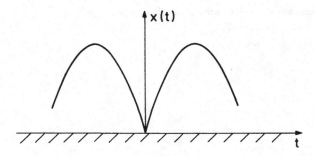

Fig. 2.1   A bouncing ball

Its velocity

$$\dot{x}(t) = \begin{cases} -gt - v_0, & t < 0 \\ -gt + v_0, & t > 0 \end{cases} \qquad (2.2)$$

is discontinuous at the time of impact.

Therefore the force felt by the ball during the impact is infinite but of infinitely short duration. For simplicity let us switch off gravity but let us look at the details of the impact more realistically. Due to the elasticity of ball and floor the force $F_b(t)$ exerted on the ball by the floor is an ordinary function and may look like in fig. 2.3.

If instead of a rubber ball we plot the force on a glass marble, we get a higher and narrower function $F_m(t)$. The corresponding velocities are sketched in fig. 2.4.

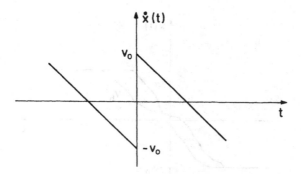

Fig. 2.2   The velocity of the idealized bouncing ball

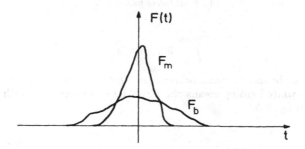

Fig. 2.3   The forces on ball and marble

Assuming energy conservation both force functions have the same integral, the momentum transfer:

$$2Mv_0 = \int_{-\infty}^{+\infty} F_b(t)dt = \int_{-\infty}^{+\infty} F_m(t)dt. \tag{2.3}$$

To recover our initial situation, equation (2.1) with $g = 0$, we consider the limit of an infinitely stiff ball

$$F_\infty(t) = \begin{cases} 0, & t \neq 0 \\ \infty, & t = 0 \end{cases}. \tag{2.4}$$

Although unrealistic physically, we want this limit to ignore the details of the impact and to describe correctly universal properties of any bouncing ball, e.g. energy conservation, and trajectories. In other words, we are looking for a universal "function".

$$\delta(x) = \begin{cases} 0, & x \neq 0 \\ \infty, & x = 0 \end{cases} \tag{2.5}$$

26

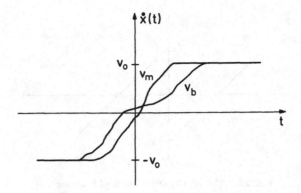

Fig. 2.4 Velocities of ball and marble

such that

$$\int_{-\infty}^{+\infty} \delta(x)dx = 1. \tag{2.6}$$

As second example consider monochromatic light of frequency ν_0 and energy E produced from white light by means of a filter. Its spectrum $\Delta(\nu)$ with band width ϵ is sketched in fig. 2.5.

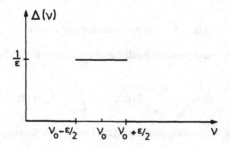

Fig. 2.5 Frequency spectrum with band width ϵ

As in the previous example the limit as ϵ tends to zero does not yield a function. However, for an observable quantity like the energy this limit does make sense:

$$E = \int_0^\infty \Delta(\nu)h\nu d\nu = \int_{\nu_0-\epsilon/2}^{\nu_0+\epsilon/2} \frac{h\nu}{\epsilon} d\nu = \frac{h}{\epsilon} \frac{(\nu_0 + \epsilon/2)^2 - (\nu_0 - \epsilon/2)^2}{2} \xrightarrow[\epsilon\to 0]{} h\nu_0 \tag{2.7}$$

suggesting to define the ideal monochromatic spectrum by the universal "function" $\delta(x - \xi)$ satisfying the so-called sifting property:

$$\int_{-\infty}^{+\infty} \delta(x - \xi)f(x)dx = f(\xi). \tag{2.8}$$

Limits similar to these examples appear everywhere in physics. In some cases they are purely technical and unphysical, e.g. the idealization in the first example contradicts special relativity. In other cases the physical reality of such limits is a question of belief like the existence of pointlike particles or the renormalization in quantum field theory. Kirchhoff (1882) was the first one to define the universal function $\delta(x)$ by equations (2.5) and (2.8). He used its three-dimensional generalization in wave mechanics to give a precise formulation of Huygens' principle. It was also applied to electrodynamics by Heaviside (1893, 1894). He realized that $\delta(x)$ could in some sense be interpreted as derivative of the so-called Heaviside or step function

$$H(x) = \begin{cases} 0 & , \quad x < 0 \\ 1 & , \quad x > 0 \end{cases}, \tag{2.9}$$

compare with fig. 2.4. Finally Dirac (1930) popularized the use of $\delta(x)$ in the context of quantum mechanics and today $\delta(x)$ is called Dirac's δ-function or Dirac distribution. For a historical account of distributions the reader is referred to Temple (1953).

Distributions are the generalization of functions that allow for the limits in the above examples. Distributions are related to functions as real numbers are related to rational numbers. To account for experimental physics rational numbers are perfectly sufficient and depending on the desired accuracy $\sqrt{2}$ may be represented by 1.41, 1.4142 or 1.4141. The same is true for distributions. The δ distribution can be represented by one of the functions of fig. 2.3 or fig. 2.5. In order to explain the accuracy of the representation we have to define a notion of convergence, that of a Cauchy sequence for rational numbers, and the so-called weak convergence for sequences of functions. The real number $\sqrt{2}$ then is an equivalence class of sequences of rational numbers. This definition gets rid of all arbitrariness involved in the representation of $\sqrt{2}$ by rational numbers. Similarly, the definition of a distribution reflects those of its properties that do not depend on any particular choice of function used for its representation. As for real numbers there is an alternative axiomatic approach to distributions, i.e. the functional definition introduced by Laurent Schwartz (1950) which is convenient for many abstract purposes. However, we shall stick to Temple's definition (1955) of a distribution in terms of weakly convergent sequences of functions as it is tailored to the physicist's needs.

2.2 Weak convergence

For simplicity we first consider only real functions of one real variable. We shall see later that the generalization to functions over N-dimensional real space

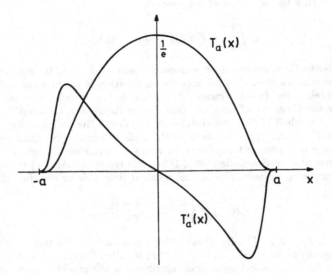

Fig. 2.6 The test function T_a and its first derivative

with complex values is straightforward. To define weak convergence, we need an important class of auxiliary functions, the test functions. A test function $T : \mathbb{R} \to \mathbb{R}$ is a differentiable function with compact support. From now on differentiable will always mean infinitely many times differentiable. Recall that the support of a function is by definition the adherence of the set of all points $x \in \mathbb{R}$ such that $T(x)$ is non-zero. Therefore a test function vanishes outside a finite interval. Note that no non-vanishing test function is analytic. A typical example of a test function is

$$T_a(x) := \begin{cases} e^{-a^2/(a^2-x^2)}, & |x| < a \\ 0, & |x| \geq a \end{cases} \tag{2.10}$$

where $a > 0$. Its support has length $2a$ and it is differentiable, for instance the first derivative of T_a

$$T_a'(x) = \begin{cases} -\frac{a^2 2x}{(a^2-x^2)^2} e^{-a^2/(a^2-x^2)}, & |x| < a \\ 0, & |x| \geq a \end{cases} \tag{2.11}$$

is also well defined everywhere:

$$\lim_{x \to a+} T_a'(x) = 0 = \lim_{x \to a-} T_a'(x). \tag{2.12}$$

Definition: A sequence of differentiable functions $f_n : \mathbb{R} \to \mathbb{R}$ $n = 1, 2, 3, \ldots$ is said to be weakly convergent if for any test function $T(x)$ the limit of numbers

$$\lim_{n \to \infty} \int_{-\infty}^{+\infty} f_n(x)T(x)dx \tag{2.13}$$

exists.

Example: The sequence of functions $f_n(x) := (n/\pi)1/(1 + n^2 x^2)$ converges weakly and we have

$$\lim_{n \to \infty} \int_{-\infty}^{+\infty} f_n(x)T(x)dx = T(0), \tag{2.14}$$

i.e. the sifting property.

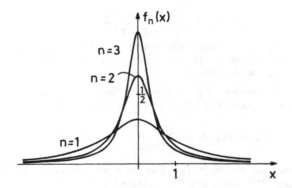

Fig. 2.7   The sequence of Breit-Wigner functions $f_n(x) = \frac{n}{\pi}\frac{1}{1+n^2x^2}$

Proof: Let $T(x)$ be a test function. Every test function is bounded: $|T(x)| < B$ for some real number B independent of x. We split the integral into three pieces:

$$\int_{-\infty}^{+\infty} \frac{n}{\pi}\frac{1}{1+n^2x^2}T(x)dx = \int_{-\infty}^{-1/\sqrt{n}} + \int_{-1/\sqrt{n}}^{1/\sqrt{n}} + \int_{1/\sqrt{n}}^{\infty} \frac{n}{\pi}\frac{1}{1+n^2x^2}T(x)dx. \tag{2.15}$$

The first and third piece tend to zero for large n.

$$\left| \int_{1/\sqrt{n}}^{\infty} \frac{n}{\pi}\frac{1}{1+n^2x^2}T(x)dx \right| \leq B \int_{1/\sqrt{n}}^{\infty} \frac{n}{\pi}\frac{1}{1+n^2x^2}dx$$

$$= B(\frac{1}{2} - \frac{1}{\pi}\arctan\sqrt{n}) \xrightarrow[n \to \infty]{} 0. \tag{2.16}$$

To compute the second piece we use a mean value theorem: There is a number ξ in the domain of integration $\xi \in [-1/\sqrt{n}, 1/\sqrt{n}]$ such that

$$\int_{-1/\sqrt{n}}^{1/\sqrt{n}} \frac{n}{\pi} \frac{1}{1+n^2x^2} T(x)dx = T(\xi) \int_{-1/\sqrt{n}}^{1/\sqrt{n}} \frac{n}{\pi} \frac{1}{1+n^2x^2} dx$$

$$= T(\xi)\frac{2}{\pi} \arctan \sqrt{n} \xrightarrow[n \to \infty]{} T(0). \tag{2.17}$$

In words: Since the sequence of Breit-Wigner functions peaks more and more at the origin, it filters out the values of the test function there.

It should be clear from this picture that there are many other sequences of functions that have this property. For instance

$$f_n(x) = \frac{n}{\sqrt{\pi}} e^{-n^2x^2} \tag{2.18}$$

and

$$f_n(x) = \frac{1}{n\pi} \frac{\sin^2 nx}{x^2} \tag{2.19}$$

are weakly convergent sequences which also have the sifting property, equation (2.14). Any of these sequences (or any function f_n with sufficiently large index n) represents the universal function δ.

Definition: A distribution D (or generalized function) is an equivalence class of weakly convergent sequences of functions $[f_n]$, and we write

$$\int D(x)T(x)dx := \lim_{n \to \infty} \int_{-\infty}^{+\infty} f_n(x)T(x)dx \in \mathbb{R} \tag{2.20}$$

for any representative sequence f_n. The lhs is also denoted by $\int DT$ or in the functional setting $D(T)$ and reads "D evaluated on T".

Indeed a distribution is a linear functional over the space of test functions:

$$D(T+S) = D(T) + D(S) \tag{2.21}$$

for two test functions T and S, and

$$D(aT) = aD(T) \tag{2.22}$$

where a is a real number.

Two weakly convergent sequences of functions $f_n(x)$ and $g_n(x)$ are equivalent if their difference converges weakly to zero. Note that weak convergence does not imply pointwise convergence, e.g. the limit of real numbers obtained from the weakly convergent sequence in our example $f_n(x) = (n/\pi)1/(1+n^2x^2)$ by putting $x = 0$ fails to exist.

$$\lim_{n \to \infty} f_n(0) = \lim_{n \to \infty} \frac{n}{\pi} = \infty.$$

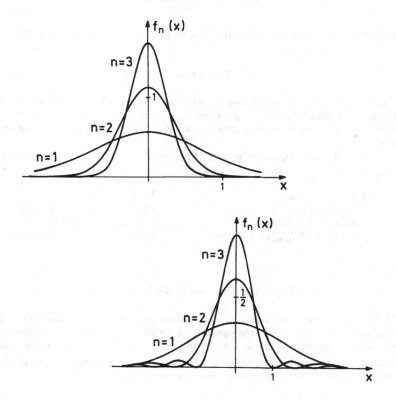

Fig. 2.8   Two other representations of the δ-distribution

$$f_n(x) = \frac{n}{\sqrt{\pi}} e^{-n^2 x^2} \quad \text{and} \quad f_n(x) = \frac{1}{n\pi} \frac{\sin^2 nx}{x^2}$$

Therefore it makes no sense to talk about the value of a distribution $D(x)$ at a point x. A distribution can only be "localized" in a finite interval by evaluating it on a test function with support in that interval, "the distribution is smeared out by the test function". This already suggests that distributions are better suited to formulate the uncertainty principle in quantum field theory than functions.

We can now define the most prominent example of a distribution.

32

Definition: The distribution represented by

$$f_n(x) = \frac{n}{\pi} \frac{1}{1 + n^2(x - \xi)^2} \tag{2.23}$$

is called (Dirac's) delta distribution at $\xi \in \mathbb{R}$ and denoted by δ_ξ or $\delta_\xi(x)$ or $\delta(x - \xi)$.

Exercises

1. Find a pointwise convergent sequence of functions $f_n(x)$ (i.e. for every point x the limit of numbers $\lim_{n \to \infty} f_n(x)$ exists) which does not converge weakly.
2. Show that the two sequences defined by equations (2.18) and (2.19) represent $\delta(x)$.
3. Are the following sequences weakly convergent? If so, what distributions do they define?

   a) $f_n(x) = 1/nx$,   b) $f_n(x) = x/n$,
   c) $f_n(x) = T_{1/2}(x - n)$,   d) $f_n(x) = n\,T_{1/2}(x - n)$,
   e) $f_n(x) = 1/nT_n(x)$,   f) $f_n(x) = nT_{1/n}(x)$,
   g) $f_n(x) = n^2 T_{1/n}(x)$,   h) $f_n(x) = \frac{1}{2} + \frac{1}{\pi} \arctan nx$.

2.3 Operations with distributions

With respect to many operations distributions behave like functions. This is immediately true for the linear operations, addition and scalar multiplication, because they commute with integral and limit in the definition of weak convergence.

Addition: If $f_n(x)$ and $g_n(x)$ are two weakly convergent sequences of functions, then their sum also converges weakly. For any test function $T(x)$

$$\lim_{n \to \infty} \int (f_n + g_n)T = \lim_{n \to \infty} \left(\int f_n T + \int g_n T \right) = \lim_{n \to \infty} \int f_n T + \lim_{n \to \infty} \int g_n T. \tag{2.24}$$

Also the sum of two sequences that converge weakly to zero converges weakly to zero. Therefore the sum passes to equivalence classes. If D is represented by $f_n(x)$ and E by $g_n(x)$ then $D + E$ is represented by $(f_n + g_n)(x)$ and for any test function T

$$(D + E)(T) = D(T) + E(T). \tag{2.25}$$

Scalar multiplication: Similarly the scalar multiple of a distribution aD, $a \in \mathbb{R}$ is well defined and represented by $af_n(x)$,

$$(aD)(T) = aD(T). \tag{2.26}$$

Since there is also the distribution 0, all distributions form an infinite dimensional vector space, just as functions do.

Differentiation: Let $f_n(x)$ be a weakly convergent sequence representing a distribtution D. By hypothesis all functions $f_n(x)$ are differentiable. The sequence of their derivatives $f'_n(x)$ is again weakly convergent:

$$\lim_{n \to \infty} \int f'_n T = \lim_{n \to \infty} \left(f_n T \Big|_{-\infty}^{+\infty} - \int f_n T' \right) = -D(T'). \tag{2.27}$$

The boundary term in the integration by parts vanishes because the test function T has compact support. The second term has a well defined limit because the derivative of any test function is also a test function. The sequence $f'_n(x)$ defines a distribution D', the derivative of D. Indeed this definition of D' does not depend on the particular choice f_n used to represent D.

$$\int D'T = - \int DT'. \tag{2.28}$$

The derivative of the δ-distribution, for example, is represented in fig. 2.9.

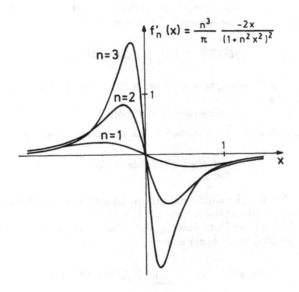

Fig. 2.9 A representation of δ'

In electrodynamics it describes the ideal dipole, a pair of opposite charges $+q$ and $-q$ separated by a distance d in the limit of d tending to zero and q to infinity such that their product qd remains constant. Likewise higher multipoles are described by higher derivatives. Here the universality expressed by the equivalence relation in the definition of a distribution means that multipoles only describe the electromagnetic field asymptotically far away from the charge "distribution".

Multiplication: We now come to a most important difference between functions and distributions: the product of two distributions is, in general, not defined. The first idea that comes to mind, defining the product by multiplying representative functions, does not work, e.g. for the square of the δ-distribution. The sequence

$$g_n(x) = \frac{n^2}{\pi^2} \frac{1}{(1 + n^2 x^2)^2} \tag{2.29}$$

fails to converge weakly. Said differently, while functions naturally form a ring, distributions do not. This has many consequences. We shall see that distributions provide a practical tool for solving linear differential equations, a tool that does not apply to non-linear equations. Einstein's and Yang-Mills equations are non-linear and the concept of point particles in general relativity or a Yang-Mills theory raises difficult problems even on a purely technical level. The renormalization programme of quantum field theory can be viewed as an attempt to define a product of distributions.

On the other hand, the product of a differentiable function $g(x)$ and a distribution D represented by the sequence $f_n(x)$ is well defined:

$$(gD)(T) := \lim_{n \to \infty} \int g f_n T = D(gT). \tag{2.30}$$

For if $T(x)$ is a test function, then so is $(gT)(x)$. Applying this construction to the δ-distribution, we get

$$g(x)\delta(x - \xi) = g(\xi)\delta(x - \xi) \tag{2.31}$$

an equation that should not be confused with the sifting property

$$\int \delta(x - \xi)T(x)dx = T(\xi). \tag{2.32}$$

Substitution: Let $h(x)$ be a diffeomorphism. (A diffeomorphism is a differentiable function admitting a differentiable inverse). If $f_n(x)$ is a weakly convergent sequence representing a distribution $D(x)$, then $f_n(h(x))$ also converges weakly and represents a distribution denoted by $D(h(x))$ and

$$
\begin{aligned}
\int D(h(x))T(x)dx &:= \lim_{n \to \infty} \int f_n(h(x))T(x)dx \\
&= \lim_{n \to \infty} \int f_n(u) \frac{T(h^{-1}(u))}{|h'(h^{-1}(u))|} du.
\end{aligned}
\tag{2.33}
$$

For example:

$$\delta(ax + b) = \frac{1}{|a|}\delta(x + b/a), \quad a \neq 0. \tag{2.34}$$

As for functions, the support of a distribution is defined to be the complement of the largest open subset on which the distribution vanishes. Since the support of the δ-distribution is concentrated in one point, it admits more general substitutions, for example $\delta(x^2 - a^2)$, for non-vanishing a. Indeed if $f_n(x)$ represents $\delta(x)$, then also $f_n(x^2 - a^2)$ converges weakly:

$$\lim_{n \to \infty} \int_{-\infty}^{+\infty} f_n(x^2 - a^2)T(x)dx = \lim_{n \to \infty} \left(\int_{-\infty}^{0} + \int_{0}^{\infty} \right) f_n(x^2 - a^2)T(x)dx$$

$$= \lim_{n \to \infty} \int_{\infty}^{-a^2} f_n(u)T(-\sqrt{u + a^2})\frac{du}{-2\sqrt{u + a^2}}$$

$$+ \lim_{n \to \infty} \int_{-a^2}^{\infty} f_n(u)T(+\sqrt{u + a^2})\frac{du}{2\sqrt{u + a^2}} \tag{2.35}$$

$$= \lim_{n \to \infty} \int_{-a^2}^{\infty} f_n(u)\frac{T(-\sqrt{u + a^2}) + T(\sqrt{u + a^2})}{2\sqrt{u + a^2}}du$$

where we have substituted $u := x^2 - a^2$, $x = \pm\sqrt{u + a^2}$. Now $\frac{T(-\sqrt{u+a^2})+T(\sqrt{u+a^2})}{2\sqrt{u+a^2}}$ is not a test function, it has a possible singularity at $u = -a^2$. However, for large n, $f_n(u)$ only filters out values at $u = 0$. To avoid the singularity at $u = -a^2$, we use equation (2.31)

$$eT_{a^2/2}(u)\delta(u) = \delta(u) \tag{2.36}$$

and replace $f_n(u)$ by $f_n(u)T_{a^2/2}(u)$. In this form we can evaluate the integral:

$$\lim_{n \to \infty} \int_{-\infty}^{+\infty} f_n(x^2 - a^2)T(x)dx =$$

$$\lim_{n \to \infty} \int_{-\infty}^{+\infty} f_n(u)eT_{a^2/2}(u)\frac{T(-\sqrt{u + a^2}) + T(\sqrt{u + a^2})}{2\sqrt{u + a^2}}du \tag{2.37}$$

$$= \frac{T(-|a|) + T(|a|)}{2|a|}.$$

Therefore

$$\delta(x^2 - a^2) = \frac{1}{2|a|}[\delta(x + |a|) + \delta(x - |a|)]. \tag{2.38}$$

This procedure generalizes to functions $h(x)$ with a finite number of simple zeros $\xi_1, \xi_2, \cdots \xi_N$ and yields

$$\delta(h(x)) = \sum_{i=1}^{N} \frac{1}{|h'(\xi_i)|}\delta(x - \xi_i). \tag{2.38}$$

Exercises

1. The weakly convergent sequence

$$f_n(x) = \frac{1}{2} + \frac{1}{\pi} \arctan nx$$

represents Heaviside's step function

$$H(x) = \begin{cases} 0, & x < 0 \\ 1, & x > 0 \end{cases}$$

considered as a distribution. Show that its derivative is the δ-distribution.

2. Show that the sequence (2.29) does not converge weakly.

3. For a differentiable function $g(x)$ prove

$$g(x)\delta(x - \xi) = g(\xi)\delta(x - \xi).$$

4. Prove the Leibniz rule

$$(gD)' = g'D + gD'$$

where g is a differentiable function, D a distribution.

5. Calculate

$$g(x)(d/dx)^n \delta(x)$$

for $n = 1, 2, 3 \cdots$.

6. Verify

$$\delta(ax + b) = \frac{1}{|a|}\delta(x + b/a), \quad a \neq 0.$$

7. Calculate $\delta(x^2 - 2x)$.

2.4 Correspondence between functions and distributions

Any differentiable function $f(x)$ can be considered as a distribution by representing it by the weakly convergent sequence $f(x), f(x), f(x), \cdots$. It will be crucial also to be able to include piecewise differentiable functions in the space of distributions. The archetype of such a function is Heaviside's step function $H(x)$. As distribution it can be represented by

$$f_n(x) = \frac{1}{2} + \frac{1}{\pi} \arctan nx. \tag{2.40}$$

Indeed

$$\lim_{n \to \infty} \int_{-\infty}^{+\infty} (\frac{1}{2} + \frac{1}{\pi} \arctan nx) T(x) dx = \int_0^{\infty} T(x) dx = \int_{-\infty}^{+\infty} H(x) T(x) dx \tag{2.41}$$

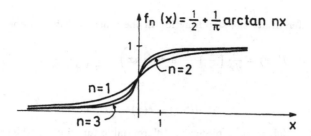

Fig. 2.10   A representation of Heaviside's step function as distribution

for any test functions T.

To prove the first equality, we split the integral into three pieces

$$\int_{-\infty}^{-1/\sqrt{n}} + \int_{-1/\sqrt{n}}^{1/\sqrt{n}} + \int_{1/\sqrt{n}}^{\infty} (\frac{1}{2} + \frac{1}{\pi} \arctan nx).$$

The first two converge to zero:

$$\left| \int_{-\infty}^{-1/\sqrt{n}} (\frac{1}{2} + \frac{1}{\pi} \arctan nx) T(x) dx \right|$$

$$\leq \int_{-\infty}^{-1/\sqrt{n}} (\frac{1}{2} + \frac{1}{\pi} \arctan nx) |T(x)| dx \qquad (2.42)$$

$$\leq (\frac{1}{2} + \frac{1}{\pi} \arctan -\sqrt{n}) \int_{-\infty}^{+\infty} |T(x)| dx \xrightarrow[n \to \infty]{} 0$$

and

$$\left| \int_{-1/\sqrt{n}}^{+1/\sqrt{n}} (\frac{1}{2} + \frac{1}{\pi} \arctan nx) T(x) dx \right|$$

$$\leq \int_{-1/\sqrt{n}}^{+1/\sqrt{n}} (\frac{1}{2} + \frac{1}{\pi} \arctan nx) |T(x)| dx \qquad (2.43)$$

$$\leq B \frac{2}{\sqrt{n}} \xrightarrow[n \to \infty]{} 0$$

where B is a bound of the test function $|T(x)| \leq B$. Finally the last piece gives the contribution indicated:

$$\left| \int_{1/\sqrt{n}}^{\infty} T(x) dx - \int_{1/\sqrt{n}}^{\infty} (\frac{1}{2} + \frac{1}{\pi} \arctan nx) T(x) dx \right|$$

$$\leq \int_{1/\sqrt{n}}^{\infty} (\frac{1}{2} - \frac{1}{\pi} \arctan nx) |T(x)| dx \qquad (2.44)$$

$$\leq (\frac{1}{2} - \frac{1}{\pi} \arctan \sqrt{n}) \int_{-\infty}^{\infty} |T(x)| dx \xrightarrow[n \to \infty]{} 0.$$

38

Now we can differentiate the step function in the sense of distributions:

$$f'_n(x) = \frac{d}{dx}\left(\frac{1}{2} + \frac{1}{\pi}\arctan nx\right) = \frac{n}{\pi}\frac{1}{1 + n^2 x^2}. \tag{2.45}$$

Therefore

$$H'(x) = \delta(x). \tag{2.46}$$

Next we want to differentiate piecewise differentiable functions in the sense of distributions.

Definition: We call a function $f(x)$ piecewise differentiable if its domain of definition can be decomposed into a finite number of closed subsets such that $f(x)$ is differentiable in the interior of each of these closed subsets and can be continued differentiably to a neighbourhood of each closed subset.

In other words, a piecewise differentiable function (in one dimension) has a finite number of finite jumps and kinks like the trajectory of the bouncing ball (2.1) or its velocity (2.2). According to our convention $f(x) = 1/x$ is not piecewise differentiable. To be specific let us consider a piecewise differentiable function with two jumps:

$$f(x) = \begin{cases} f_1(x), & x < \xi_1 \\ f_2(x), & \xi_1 < x < \xi_2 \\ f_3(x), & x > \xi_2 \end{cases} \tag{2.47}$$

with three differentiable functions $f_1(x)$, $f_2(x)$ and $f_3(x)$.

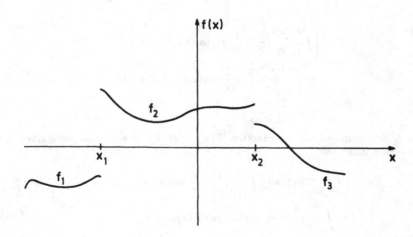

Fig. 2.11 A piecewise differentiable function

We can write $f(x)$ under the following form:

$$f(x) = f_3(x)H(x - \xi_2)$$
$$+ f_2(x)H(x - \xi_1) - f_2(x)H(x - \xi_2) \qquad (2.48)$$
$$+ f_1(x) - f_1(x)H(x - \xi_1).$$

Being a sum of products of a differentiable function with a distribution any piecewise differentiable function is a distribution and can be differentiated:

$$f'(x) = f_3'(x)H(x - \xi_2)$$
$$+ f_2'(x)H(x - \xi_1) - f_2'(x)H(x - \xi_2)$$
$$+ f_1'(x) - f_1'(x)H(x - \xi_1)$$
$$+ f_3(x)\delta(x - \xi_2) + f_2(x)\delta(x - \xi_1) - f_2(x)\delta(x - \xi_2) - f_1(x)\delta(x - \xi_1) \quad (2.49)$$
$$= \begin{cases} f_1'(x), & x < \xi_1 \\ f_2'(x), & \xi_1 < x < \xi_2 \\ f_3'(x), & x > \xi_2 \end{cases} + [f_3(\xi_2) - f_2(\xi_2)]\delta(x - \xi_2)$$
$$+ [f_2(\xi_1) - f_1(\xi_1)]\delta(x - \xi_1).$$

Here we have used the Leibniz rule for differentiating a product of a differentiable function g and a distribution D

$$(gD)' = g'D + gD' \qquad (2.50)$$

and equation (2.31). In words: the derivative of a piecewise differentiable function in the sense of distributions is the piecewise derivative in the sense of functions plus — for each jump — a δ-distribution located at the discontinuity multiplied by the amplitude of the jump together with its sign. For example, the force on the bouncing ball is

$$F(t) := M\ddot{x}(t) = -Mg + 2Mv_0\delta(t). \qquad (2.51)$$

Although piecewise differentiable functions are sufficient for the applications we have in mind, let us note that other more general functions like continuous functions are distributions and can be differentiated as such. For a continuous function one uses the following theorem to represent it by a weakly convergent sequence.

Theorem: For any continuous function $f(x)$ there is a sequence of differentiable functions $f_n(x)$ that converges to $f(x)$ uniformly in any compact subset.

The proof is by explicit construction

$$f_n(x) = \frac{\int f(\xi)T_{1/n}(x - \xi)d\xi}{\int T_{1/n}(x - \xi)d\xi}. \qquad (2.52)$$

The theorem also tells us that we loose nothing by restricting ourselves to differentiable functions in the definition of a distribution.

Exercises

1. In the sense of distributions calculate the first and second derivative of

$$f(x) = \sqrt{(x^2 - a^2)^2}$$

2. Compute the first and second derivative of

$$f(x) = \begin{cases} 0, & x < 0 \\ \frac{1}{\omega} e^{-\lambda x} \sin \omega x, & x > 0 \end{cases}$$

where ω and λ are positive constants.

2.5 Sequences of distributions

Having defined the real numbers as convergent sequences of rational numbers, it is useful to take one further step and consider convergent sequences of real numbers. The same is true for distributions.

Definition: A sequence of distributions D_n, $n = 1, 2, 3...$, is said to be convergent if there is a distribution D such that for any test function T

$$\lim_{n \to \infty} \int_{-\infty}^{+\infty} D_n T = \int_{-\infty}^{+\infty} DT. \tag{2.53}$$

We then write

$$\lim_{n \to \infty} D_n = D. \tag{2.54}$$

E.g.

$$\lim_{n \to \infty} \delta_n(x) = 0. \tag{2.55}$$

Any weakly convergent sequence of functions converges as sequence of distributions, e.g.

$$D_n(x) := \cos nx, \tag{2.56}$$

$$\lim_{n \to \infty} D_n(x) = 0 \tag{2.57}$$

because for any test function $T(x)$

$$\int_{-\infty}^{+\infty} \cos nx \, T(x) dx = \frac{\sin nx}{n} T(x) \Big|_{-\infty}^{+\infty} - \frac{1}{n} \int_{-\infty}^{+\infty} \sin nx T'(x) dx \xrightarrow[n \to \infty]{} 0. \tag{2.58}$$

Limits of distributions have some convenient properties not shared by limits of functions. There are several notions of convergence for sequences of functions, only one for distributions, and a limit of distributions needs no further specification. In the case of distributions integration and limit can be interchanged

$$\lim_{n\to\infty} \int D_n(x)T(x)dx = \int \lim_{n\to\infty} D_n(x)T(x)dx. \tag{2.59}$$

This is just the definition of the limit. Also differentiation and limit can be interchanged

$$\lim_{n\to\infty} \frac{d}{dx}D_n(x) = \frac{d}{dx}\lim_{n\to\infty} D_n(x). \tag{2.60}$$

Indeed for any test function $T(x)$ we have

$$\lim_{n\to\infty} \int D'_n T = \lim_{n\to\infty} \left(-\int D_n T'\right) = -\int DT' = \int D'T. \tag{2.61}$$

The following example shows that such an interchange is not permissible for pointwise convergence of functions:

$$\lim_{n\to\infty} \frac{\sin nx}{n} = 0 \tag{2.62}$$

while the limit of derivatives $\lim_{n\to\infty} \cos nx$ fails to exist.

We close this section with a particular, most important limit of a sequence of distributions. Let $f(x)$ be a piecewise differentiable function not necessarily integrable. Then

$$f(x) = \int_{-\infty}^{+\infty} \delta_\xi(x)f(\xi)d\xi. \tag{2.63}$$

The rhs is understood as the limit in the sense of distributions of an appropriate Riemannian sum

$$f(x) = \lim_{n\to\infty} \sum_{i=1}^{n} \delta(x - \xi_i)f(\xi_i)\Delta\xi_i. \tag{2.64}$$

and is also called the convolution product of f and δ. Equation (2.63) should not be confused with the sifting equation

$$T(x) = \int_{-\infty}^{+\infty} \delta_x(\xi)T(\xi)d\xi.$$

To verify equation (2.64) we have to evaluate it on a test function $T(x)$:

$$\lim_{n\to\infty} \int_{-\infty}^{+\infty} \sum_{i=1}^{n} \delta_{\xi_i}(x)f(\xi_i)\Delta\xi_i T(x)dx = \lim_{n\to\infty} \sum_{i=1}^{n} f(\xi_i)\Delta\xi_i \int_{-\infty}^{+\infty} \delta_{\xi_i}(x)T(x)dx$$

$$= \lim_{n\to\infty} \sum_{i=1}^{n} f(\xi_i)T(\xi_i)\Delta\xi_i = \int_{-\infty}^{+\infty} f(x)T(x)dx. \tag{2.65}$$

Exercises

1. Calculate:

$$\lim_{n \to \infty} \frac{1}{n^2} \delta''(x - n^2).$$

2. Does the sequence

$$D_n = \left(\frac{d}{dx}\right)^n \delta(x)$$

converge?

3. Show that the sequence

$$f_n(x) := n f(x/n)$$

converges to the δ-distribution where $f(x)$ is a piecewise differentiable function such that

$$\int_{-\infty}^{+\infty} f(x)dx = 1.$$

4. Calculate

$$\lim_{n \to \infty} \frac{\delta(x - 1/n) - \delta(x)}{1/n}$$

2.6 Green functions

A homogeneous linear differential equation is by definition an equation of the type

$$a_N(x)f^{(N)}(x) + a_{N-1}(x)f^{(N-1)}(x) + \ldots + a_2(x)f''(x) + a_1(x)f'(x) + a_0(x)f(x) = 0 \tag{2.66}$$

with differentiable "coefficient" functions $a_i(x)$, $i = 0, 1, \ldots, N$. The unknown is the function $f(x)$, $f^{(i)}$ denotes the i-th derivative of f. If $a_N(x)$ is non-zero, $N \geq 1$, the equation is said to be of degree N. It is convenient to rewrite this equation in the form:

$$Af = 0 \tag{2.67}$$

with the linear differential operator

$$A := \sum_{j=0}^{N} a_j(x) \left(\frac{d}{dx}\right)^j. \tag{2.68}$$

The word "linear" is used because any linear combination (with constant coefficients) of solutions is again a solution:

$$A(f + g) = Af + Ag = 0 \tag{2.69}$$

$$A(Kf) = KAf = 0, \quad k \in \mathbb{C}. \tag{2.70}$$

This is what physicists call superposition principle motivated by the wave equation which is linear. In more mathematical terms: the set of all solutions to a linear differential equation forms a vector space. An inhomogeneous linear differential equation is of the form

$$Af(x) = s(x) \tag{2.71}$$

with a given "source" function $s(x)$.

Since distributions can be differentiated and multiplied by differentiable functions, a linear differential equation also makes sense for distributions

$$AD(x) = s(x) \tag{2.72}$$

if the coefficients $a_i(x)$ are differentiable functions and the source is a distribution. Consider for example the equation

$$xf'(x) = 0. \tag{2.73}$$

In the sense of functions its most general solution is a constant K_1. In the sense of distributions there is a second independent solution

$$D(x) = K_2 H(x). \tag{2.74}$$

Of course there are linear differential equations for distributions that do not admit any non-trivial solution, e.g.

$$(x^2 \frac{d}{dx} + 1)D(x) = 0. \tag{2.75}$$

Its most general solution in the sense of functions

$$f(x) = K e^{1/x} \tag{2.76}$$

cannot be represented by a weakly convergent sequence of functions if K is non-zero.

Definition: A Green function (or propagator or elementary solution) of a linear differential operator

$$A = \sum_{j=0}^{N} a_j(x) \left(\frac{d}{dx} \right)^j \tag{2.77}$$

with differentiable coefficient functions $a_i(x)$ is a distribution $G_\xi(x)$ such that

$$AG_\xi(x) = \delta_\xi(x). \tag{2.78}$$

Despite its name the Green function is not a function in general. However, in many important examples the Green function turns out to be a function and this terminology has become so popular that we cannot refrain from using it. The reader should also be warned that some authors insert an additional minus sign or a complex i on the rhs of equation (2.78).

Knowledge of the Green functions for all $\xi \in \mathbb{R}$ is extremely valuable, because they can solve the inhomogeneous equation for many source functions $s(x)$ by means of the magic formula

$$D(x) := \int_{-\infty}^{+\infty} G_\xi(x)s(\xi)d\xi. \tag{2.79}$$

Indeed, if this integral understood as a limit of distributions of a corresponding Riemannian sum exists, then it satisfies the inhomogeneous equation:

$$AD(x) = A \int_{-\infty}^{+\infty} G_\xi(x)s(\xi)d\xi = A \lim_{n\to\infty} \sum_{i=1}^{n} G_{\xi_i}(x)s(\xi_i)\Delta\xi_i$$

$$= \lim_{n\to\infty} \sum_{i=1}^{n} AG_{\xi_i}(x)s(\xi_i)\Delta\xi_i = \lim_{n\to\infty} \sum_{i=1}^{n} \delta_{\xi_i}(x)s(\xi_i)\Delta\xi_i \tag{2.80}$$

$$= \int_{-\infty}^{+\infty} \delta_\xi(x)s(\xi)d\xi = s(x).$$

Example: Consider

$$A = \frac{d}{dx} \tag{2.81}$$

with Green functions

$$G_\xi(x) = H(x - \xi). \tag{2.82}$$

Then the magic formula even makes sense as equation for functions:

$$f(x) = \int_{-\infty}^{+\infty} H(x - \xi)s(\xi)d\xi = \int_{-\infty}^{x} s(\xi)d\xi. \tag{2.83}$$

The integral exists if the source function is integrable at $-\infty$ and we have recovered the fundamental theorem of differentiation and integration. We also see that the Green function is not unique, $H(x-\xi)+K$ is also a Green function for any constant K. This ambiguity can be avoided by choosing an initial condition for instance at $x = -\infty$. This choice then dictates what source functions may be used, for $K = 0$ $s(x)$ must be integrable at $-\infty$, for $K = -1$ at $+\infty$. For all other K's the source must be integrable at both ends.

If the coefficients of the linear differential operator A are all constant, translation invariance implies that the Green function is of the form

$$G_\xi(x) = G(x - \xi). \tag{2.84}$$

like in the example. The magic formula now reduces to a convolution product and its benefits are accessible if we can solve the one inhomogeneous equation

$$AG(x) = \delta(x). \tag{2.85}$$

Consider the damped harmonic oscillator subject to an external force:

$$\left(\left(\frac{d}{dt} \right)^2 + 2\lambda \frac{d}{dt} + \omega_0^2 \right) f(t) =: A f(t) = s(t) \tag{2.86}$$

with $\omega_0 > \lambda > 0$. The Green function $G(t)$ is the time evolution of the oscillator initially at rest after being hit at $t = 0$ by a sharp impulse. We can guess $G(t)$ from the solution of the homogeneous equation:

$$G(t) = \begin{cases} 0, & t < 0 \\ \frac{1}{\omega} e^{-\lambda t} \sin \omega t, & t > 0. \end{cases} \tag{2.87}$$

with

$$\omega := \sqrt{\omega_0^2 - \lambda^2}. \tag{2.88}$$

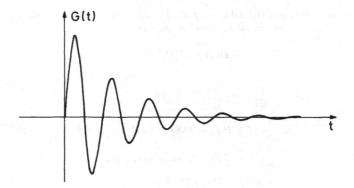

Fig. 2.12   The causal Green function of the damped harmonic oscillator

Any source $s(t)$ can be thought of as a superposition of sharp impulses

$$s(t) = \int \delta(t - \tau) s(\tau) d\tau \tag{2.89}$$

and the magic formula is just the superposition principle (the zero dimensional analogue of Huygens' principle):

$$f(t) = \int G(t - \tau) s(\tau) d\tau. \tag{2.90}$$

In this example we have a physical reason to choose the initial condition as in equation (2.87), the so-called causal Green function. We want the source, the sharp impulse at $t = 0$, to be the cause, the Green function to be the effect, the response of the system which is posterior to the cause. As expected intuitively, the integral (2.90) exists for any bounded source that sets in at a finite time t_0:

$$f(t) = \int_{t_0}^{t} \frac{1}{\omega} e^{-\lambda(t-\tau)} \sin\omega(t-\tau)s(\tau)d\tau. \tag{2.91}$$

The Green function (2.87) has a second most important application. It solves the initial value problem. I.e. given the position a and velocity b of the oscillator at $t = 0$, what is its position $f(t)$ at later times?

$$Af(t) = 0, \quad t > 0 \tag{2.92}$$

$$f(0) = a, \tag{2.93}$$

$$\dot{f}(0) = b. \tag{2.94}$$

Suppose we already know the solution $f(t)$. Let us define a distribution $D(t)$ equal to $f(t)$ for positive times, equal to zero for negative times:

$$D(t) = H(t)f(t). \tag{2.95}$$

Then

$$
\begin{aligned}
AD &= \left(\frac{d^2}{dt^2} + 2\lambda\frac{d}{dt} + \omega_0^2 \right) Hf \\
&= \frac{d}{dt}(\delta f + H\dot{f}) + 2\lambda\delta f + 2\lambda H\dot{f} + \omega_0^2 Hf \\
&= \frac{d}{dt}(a\delta + H\dot{f}) + 2\lambda a\delta + 2\lambda H\dot{f} + \omega_0^2 Hf \\
&= a\dot{\delta} + \delta\dot{f} + H\ddot{f} + 2\lambda a\delta + 2\lambda H\dot{f} + \omega_0^2 Hf \\
&= a\dot{\delta} + (2\lambda a + b)\delta + HAf = a\dot{\delta} + (2\lambda a + b)\delta.
\end{aligned}
\tag{2.96}
$$

We can solve this equation for D using the magic formula with source

$$s = a\dot{\delta} + (2\lambda a + b)\delta \tag{2.97}$$

This source is a distribution and the magic formula will involve a convolution product of two distributions. However, the distributions in the source are the δ-distribution and its derivative, and we do not need the general theory of convolutions. The convolution product of two functions $f(x)$ and $g(x)$ is by definition:

$$(f * g)(x) := \int_{-\infty}^{+\infty} f(x - \xi)g(\xi)d\xi \tag{2.98}$$

As already mentioned, this integral does not coverge for arbitrary functions, which makes the general theory complicated. It does converge for instance if one of the functions has compact support. Then it defines a product which is commutative, associative, and distributive. This convolution product generalizes to distributions and has two additional important properties:

- The δ-distribution is the neutral element

$$\delta * D = D \tag{2.99}$$

for any distribution D. Compare also with equation (2.63).

- and

$$\left(\frac{d}{dx}D\right) * E = D * \left(\frac{d}{dx}E\right) \tag{2.100}$$

for any distributions D and E such that the above convolutions exist.

In this notation the magic formula (2.79) reads

$$D = G * s \tag{2.101}$$

and for the source (2.97) we obtain:

$$\begin{aligned} D &= G * [a\dot{\delta} + (2\lambda a + b)\delta] \\ &= a\dot{G} + (2\lambda a + b)G \\ &= H(t)\left[\frac{\lambda a + b}{\omega}e^{-\lambda t}\sin\omega t + ae^{-\lambda t}\cos\omega t\right]. \end{aligned} \tag{2.102}$$

Indeed

$$f(t) = \frac{\lambda a + b}{\omega}e^{-\lambda t}\sin\omega t + ae^{-\lambda t}\cos\omega t \tag{2.103}$$

solves the initial value problem. Equation (2.102) which propagates initial values to later times also explains why the (causal) Green function is often called propagator.

So far we have derived the Green function of the harmonic oscillator by guess work. Chapter 4 contains an algorithm that permits calculating this Green function using Fourier transforms and the integration tricks of chapter 1.

Exercises

1. Compute Green functions $G_\xi(x)$ of the following differential operators

   $a)$ $x^2 + 1$, $b)$ $\left(\frac{d}{dx}\right)^n$, $n = 1, 2, 3, ...$, $c)$ $e^x\frac{d}{dx}$, $d)$ $(x^2 + 1)d^2/dx^2$.

2. Prove that equation (2.87) is a Green function of the damped harmonic oscillator.

3. Find the causal Green function of $d/dt - \lambda$, $\lambda \in \mathbb{C}$.

4. Consider a stretched string at rest under an external load $s(x)$. Its displacement $f(x)$ satisfies the equation

$$\frac{d^2 f}{dx^2}(x) = s(x)$$

with boundary condition $f(0) = f(L) = 0$ where L is the length of the string without load. Show that

$$G_\xi(x) = \begin{cases} -\frac{x(L-\xi)}{L}, & 0 \le x \le \xi \\ -\frac{\xi(L-x)}{L}, & \xi \le x \le L \end{cases}$$

is its Green function. Why is it unique? Why is it not of the form $G(x - \xi)$? Why is it symmetric

$$G_\xi(x) = G_x(\xi)?$$

What is its physical interpretation?

5. Let $f(x)$ and $g(x)$ be differentiable functions $f(x)$ with compact support. Show that

$$f * g = g * f \quad \text{and} \quad f' * g = f * g'.$$

6. Let $s(t)$ be a source of compact support. Solve the initial value problem of the inhomogeneous harmonic osciallator:

$$Af = s, \quad t > 0$$
$$f(0) = a \quad \text{and} \quad \dot{f}(0) = b.$$

7. Let $f(t)$ and $g(t)$ be two linearly independent solutions of the undampened homogeneous oscillator $d^2/dt^2 + \omega^2$. Define the time-ordered product

$$T(f(t)g(\tau)) := \begin{cases} f(t)g(\tau), & t < \tau \\ f(\tau)g(t), & \tau < t. \end{cases}$$

For what real number a is

$$G_\tau(t) = aT(f(t)g(\tau))$$

a Green function of $d^2/dt^2 + \omega^2$?

8. Consider the differential operator on the positive half axis $x > 0$

$$x\frac{d^2}{dx^2} + \frac{d}{dx} - \frac{l^2}{x}$$

with a non-negative constant l. Show that for $l = 0$

$$G_\xi(x) = \begin{cases} log\xi, & x < \xi \\ logx, & x > \xi \end{cases}$$

and for $l > 0$

$$G_\xi(x) := \begin{cases} [(x/\xi)^l - (x\xi)^l]/l, & x < \xi \\ [(\xi/x)^l - (\xi x)^l]/l, & x > \xi \end{cases}$$

for Green functions.

2.7 Distributions in higher dimensions

All definitions and theorems of this chapter generalize naturally to functions of several real variables with complex values after replacing ordinary Riemann integrals by multiple integrals and ordinary derivatives by partial derivatives. We give only a few examples.

Definition: A sequence of differentiable functions $f_n : \mathbb{R}^N \to \mathbb{C}$ is said to be weakly convergent if for any differentiable function $T : \mathbb{R}^N \to \mathbb{C}$ with compact support ("test function") the limit:

$$
\lim_{n \to \infty} \int_{-\infty}^{+\infty} \cdots \int_{-\infty}^{+\infty} f_n(x^1, x^2, ..., x^N) T(x^1, x^2, ..., x^N) dx^1 dx^2 ... dx^N
$$
$$
=: \lim_{n \to \infty} \int_{\mathbb{R}^N} f_n(x) T(x) dx =: \lim_{n \to \infty} \int_{\mathbb{R}^N} f_n T
$$

(2.104)

exists where we have put $x := (x^1, x^2, ..., x^N)$.

A typical test function is

$$
T_a(x) := \begin{cases} e^{-a^2/(a^2 - r^2)}, & r < a \\ 0, & r \geq a \end{cases}
$$

(2.105)

with $a > 0$ and

$$
r := |x| := \sqrt{(x^1)^2 + (x^2)^2 + ... + (x^N)^2}.
$$

(2.106)

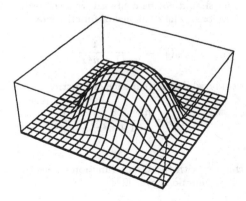

Fig. 2.13   The test function $T_a(x)$ in two dimensions

Theorem: Let $D(x)$ be a distribution represented by the weakly convergent sequence $f_n(x)$. Then $\partial/\partial x^i f_n(x)$, $i = 1, 2, ... N$ is also weakly convergent and represents a distribution denoted by $\partial/\partial x^i D(x)$.

In addition to the operations introduced so far there is a further operation for distributions in higher dimensions, the tensor product.

Theorem: Let $D(x)$ and $E(y)$ be two distributions defined over $\mathbb{R}^N$ and $\mathbb{R}^M$, represented by two weakly convergent sequences $f_n(x)$ and $g_n(y)$. Then $f_1(x)g_1(y)$, $f_2(x)g_2(y)$, $f_3(x)g_3(y)$, ... is a weakly convergent sequence of functions over $\mathbb{R}^{N+M}$ and defines a distribution denoted by $D(x)D(y)$ or $(D \otimes E)(x, y)$.

The proof relies on the fact that any test function $T(x, y)$ over $\mathbb{R}^{N+M}$ is a uniform limit of linear combinations of functions $R_j(x)S_j(y)$, $j = 1, 2, 3, ...$ where all $R_j(x)$ and $S_j(y)$ are test functions over $\mathbb{R}^N$ and $\mathbb{R}^M$ with all supports contained in two fixed compact subsets of $\mathbb{R}^N$ and $\mathbb{R}^M$.

For example the N-dimensional δ-distribution is the tensor product of N one-dimensional δ-distributions:

Definition: The distribution represented by the weakly convergent sequence

$$f_n(x) = \prod_{i=1}^{N} \frac{n}{\pi} \frac{1}{1 + n^2(x^i - \xi^i)^2} \tag{2.107}$$

is called δ-distribution at $\xi \in \mathbb{R}^N$ and is denoted by $\delta_\xi(x)$, $\delta(x - \xi)$ or

$$\delta_{\xi^1}(x^1)\delta_{\xi^2}(x^2)...\delta_{\xi^N}(x^N) = \prod_{i=1}^{N} \delta_{\xi^i}(x^i). \tag{2.108}$$

The tensor product should not be confused with the ordinary product which is not defined for distributions: The sequence of functions over $\mathbb{R}$

$$f_n(x) = \frac{n^2}{\pi^2} \frac{1}{(1 + n^2 x^2)^2} \tag{2.109}$$

does not converge weakly while the sequence of functions over $\mathbb{R}^2$

$$f_n(x, y) = \frac{n^2}{\pi^2} \frac{1}{1 + n^2 x^2} \frac{1}{1 + n^2 y^2} \tag{2.110}$$

does converge weakly.

We close this chapter with the Green function of electrostatics. Consider differentiable vector-valued functions over $\mathbb{R}^3$:

$$E : \mathbb{R}^3 \longrightarrow \mathbb{R}^3$$
$$x = (x^1, x^2, x^3) \mapsto E(x) = (E^1(x), E^2(x), E^3(x))$$

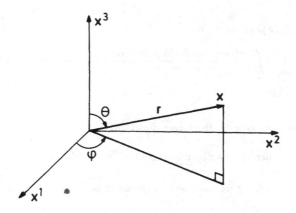

Fig. 2.14 Polar coordinates in $\mathbb{R}^3$

and the linear partial differential operator "divergence"

$$\operatorname{div} E(x) := \sum_{i=1}^{3} \frac{\partial}{\partial x^i} E^i(x). \tag{2.111}$$

It is an operator with constant coefficients and its Green function

$$G_\xi(x) = G(x - \xi) = (G^1(x - \xi), G^2(x - \xi), G^3(x - \xi)) \tag{2.112}$$

is a solution of the inhomogeneous equation

$$\operatorname{div} G(x) = \delta(x). \tag{2.113}$$

It is unique if we impose the boundary condition that $G(x)$ approaches zero as a distribution far away from the origin:

$$G(x) = \frac{1}{4\pi r^2} \frac{x}{r}. \tag{2.114}$$

This function although not piecewise differentiable is a distribution which can be seen by writing it as limit of the sequence of distributions

$$G_n(x) := \frac{x}{4\pi r^3} H(r - \frac{1}{n}). \tag{2.115}$$

To show its convergence, we introduce polar coordinates:

$$x^1 = r \cos\phi \sin\theta, \qquad r \in [0, \infty), \tag{2.116}$$

$$x^2 = r \sin\phi \sin\theta, \qquad \phi \in [0, 2\pi), \tag{2.117}$$

$$x^3 = r \cos\theta, \qquad \theta \in [0, \pi). \tag{2.118}$$

For any test function $T : \mathbb{R}^3 \to \mathbb{R}$ we have

$$\lim_{n \to \infty} \int_{\mathbb{R}^3} G_n(x)T(x)dx =$$
$$= \lim_{n \to \infty} \int_{1/n}^{\infty} \int_0^{2\pi} \int_0^{\pi} \frac{r(\cos\phi\sin\theta, \sin\phi\sin\theta, \cos\theta)}{4\pi r^3} T(x)r^2 \sin\theta d\theta d\phi dr \qquad (2.119)$$
$$= \int G(x)T(x)dx.$$

Now let us prove that $G(x)$ is the Green function.

$$\mathrm{div}G(x) = \mathrm{div}\lim_{n \to \infty} G_n(x) = \lim_{n \to \infty} \mathrm{div}G_n(x). \qquad (2.120)$$

To evaluate this last limit we use the divergence in polar coordinates for $r > 0$:

$$\mathrm{div}E = \frac{1}{r^2}\frac{\partial}{\partial r}(r^2 E_r) + \frac{1}{r\sin\theta}\frac{\partial}{\partial\phi}E_\phi + \frac{1}{r\sin\theta}\frac{\partial}{\partial\theta}(\sin\theta E_\theta) \qquad (2.121)$$

with

$$E_r := \sin\theta\cos\phi E^1 + \sin\theta\sin\phi E^2 + \cos\theta E^3, \qquad (2.122)$$
$$E_\phi := -\sin\phi E^1 + \cos\phi E^2, \qquad (2.123)$$
$$E_\theta := \cos\theta\cos\phi E^1 + \cos\theta\sin\phi E^2 - \sin\theta E^3. \qquad (2.124)$$

Therefore

$$\mathrm{div}\frac{x}{4\pi r^3} = \frac{1}{r^2}\frac{\partial}{\partial r}r^2\frac{1}{4\pi r^2} = 0, \quad r > 0. \qquad (2.125)$$

We also need the gradient of a function $f : \mathbb{R}^3 \to \mathbb{R}$

$$\mathrm{grad}f := \left(\frac{\partial}{\partial x^1}f, \ \frac{\partial}{\partial x^2}f, \ \frac{\partial}{\partial x^3}f\right) \qquad (2.126)$$

in polar coordinates.

$$(\mathrm{grad}f)_r = \frac{\partial f}{\partial r}, \qquad (2.127)$$

$$(\mathrm{grad}f)_\phi = \frac{1}{r\sin\theta}\frac{\partial f}{\partial\phi}, \qquad (2.128)$$

$$(\mathrm{grad}f)_\theta = \frac{1}{r}\frac{\partial f}{\partial\theta}, \qquad (2.129)$$

and the Leibniz rule:

$$\mathrm{div}(fE) = (\mathrm{grad}f)\cdot E + f\mathrm{div}E \qquad (2.130)$$

where the dot $\cdot$ denotes the scalar product. Now for any test function T we have:

$$
\begin{aligned}
&\lim_{n\to\infty} \int_{\mathbb{R}^3} (\mathrm{div} G_n(x)) T(x) dx \\
&= \lim_{n\to\infty} \int_{\mathbb{R}^3} \mathrm{div}(H(r-\frac{1}{n})\frac{x}{4\pi r^3}) T(x) dx \\
&= \lim_{n\to\infty} \int_{\mathbb{R}^3} \left[\mathrm{grad} H(r-\frac{1}{n}) \cdot \frac{x}{4\pi r^3} + H(r-\frac{1}{n}) \mathrm{div} \frac{x}{4\pi r^3} \right] T(x) dx \\
&= \lim_{n\to\infty} \int_0^\infty \int_0^{2\pi} \int_0^\pi \delta(r-\frac{1}{n}) \frac{x}{r} \cdot \frac{x}{4\pi r^3} T(x) r^2 \sin\theta d\theta d\phi dr \\
&= \lim_{n\to\infty} \int_0^{2\pi} \int_0^\pi T(\frac{1}{n}\cos\phi\sin\theta, \frac{1}{n}\sin\phi\sin\theta, \frac{1}{n}\cos\theta) \frac{\sin\theta}{4\pi} d\theta d\phi \\
&= T(0) = \int_{\mathbb{R}^3} \delta(x) T(x) dx.
\end{aligned}
\tag{2.131}
$$

Note that the Green function is odd

$$
G(-x) = -G(x) \tag{2.132}
$$

because the divergence has the same property and the δ-distribution is even

$$
\delta(-x) = \delta(x). \tag{2.133}
$$

The general inhomogeneous equation

$$
\mathrm{div} E(x) = \rho(x) \tag{2.134}
$$

is now solved by the magic formula

$$
E(x) = \int_{\mathbb{R}^3} \frac{\rho(\xi)}{4\pi |x-\xi|^2} \frac{x-\xi}{|x-\xi|} d\xi \tag{2.135}
$$

which is the limit of superpositions of electric fields generated by point-like charges.

Exercises

1. Are the following sequences of functions over $\mathbb{R}^2$ weakly convergent? If so, what distributions do they define?

$$
a) \ f_n(x,y) = \frac{x}{n} + \frac{n}{y}
$$

$$
b) \ f_n(x,y) = \frac{1}{1+n^2 x^2} + \arctan ny
$$

$$
c) \ f_n(x,y) = \frac{\arctan nx}{1+n^2 y^2}
$$

$$
d) \ f_n(x,y) = e^{(x^2+y^2)/n}
$$

2. Do the following functions $f : \mathbb{R}^N \to \mathbb{R}$ define distributions?

$$a) \quad \prod_{i=1}^{N} H(x^i - \xi^i) \qquad \xi \in \mathbb{R}^N$$

$$b) \quad \frac{1}{r^m}, \quad m = 1, 2, 3, \ldots$$

3. Show that for any distribution $D(x)$ over $\mathbb{R}^N$ $\partial/\partial x^i$ commutes with $\partial/\partial x^j$:

$$\frac{\partial}{\partial x^i} \frac{\partial}{\partial x^j} D(x) = \frac{\partial}{\partial x^j} \frac{\partial}{\partial x^i} D(x), \qquad i, j = 1, 2, \ldots, N.$$

4. Find a Green function for the gradient and for the curl

$$\mathrm{curl} E(x) := \left(\frac{\partial E^3}{\partial x^2} - \frac{\partial E^2}{\partial x^3}, \frac{\partial E^1}{\partial x^3} - \frac{\partial E^3}{\partial x^1}, \frac{\partial E^2}{\partial x^1} - \frac{\partial E^1}{\partial x^2} \right).$$

Hint: Use the magnetic field of the Dirac monopole.

3
Fourier series

3.1 Periodic functions

Periodic functions play an outstanding role in physics, rhythm is the secret.

Definition: A function of one real variable $f(t)$ is called periodic with period T if

$$f(t + T) = f(T) \tag{3.1}$$

for all $t \in \mathbb{R}$.

Examples are the constant functions or $\sin \frac{2\pi}{T}t$, $\cos \frac{2\pi}{T}t$, $\sin \frac{4\pi}{T}t$, $\cos \frac{4\pi}{T}t$, etc. the higher octaves. We shall refer to these functions as harmonics because they satisfy the equation of the harmonic oscillator $\ddot{f} + \omega^2 f = 0$ with $\omega := 2\pi/T$, $4\pi/T$ etc. Otherwise this word is misleading, harmonics are not harmonic functions, i.e. they do not satisfy a Laplace equation, nor do they sound harmonic when made audible. We shall often think of a periodic function as being defined on a circle of circumference T rather than on the entire real axis. Periodicity is then reproduced by wrapping the real axis around the circle.

The central question of this chapter will be: given a periodic function $f(t)$ with period T, in what sense can it be expressed as a superposition of harmonics:

$$f(t) = \sum_{j=0}^{\infty} a_j \cos \left(j \frac{2\pi}{T} t \right) + \sum_{j=1}^{\infty} b_j \sin j \frac{2\pi}{T} t? \tag{3.2}$$

The convergence of this sequence of functions, the Fourier series, will be uniform, pointwise, weak or in the mean depending on properties of the periodic function $f(t)$. One of the applications of the Fourier series is that differential equations for periodic functions are converted into algebraic equations for their Fourier coefficients a_j and b_j. For instance to see that there are no non-trivial periodic solutions of the equation

$$\left(\frac{d}{dt} + 1 \right) f = 0 \tag{3.3}$$

we plug in the Fourier series, exchange differentiation and limit, and obtain.

$$a_0 + \sum_{j=1}^{\infty} \left(j \frac{2\pi}{T} b_j + a_j \right) \cos j \frac{2\pi}{T} t + \sum_{j=1}^{\infty} \left(-j \frac{2\pi}{T} a_j + b_j \right) \sin j \frac{2\pi}{T} t = 0. \tag{3.4}$$

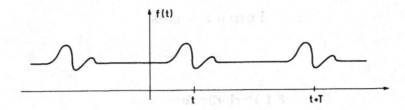

Fig. 3.1   A periodic function of period T represented on the real line and on the circle

By linear independence of the harmonics all Fourier coefficients vanish.

Similarly one finds that the only periodic solutions of

$$\left(\frac{d^2}{dt^2} + 1\right)f = 0 \tag{3.5}$$

have period $T = 2\pi$. The Fourier series reduces differential to algebraic equations essentially because the harmonics form a basis in the space of periodic functions such that each basic vector is eigenvector of the operator d^2/dt^2. This point of view will be developed in chapter 6, dealing with linear algebra in infinite dimensions.

Exercise

1. What are the eigenvalues of the harmonics with respect to d^2/dt^2?

3.2 Sequences of functions, different notions of convergence

Let $f_n(x), n = 1, 2, 3, \ldots$ be a sequence of functions defined on a convenient subset Ω of $\mathbb{R}^N$ and with complex values. For the applications we have in mind Ω will be the entire $\mathbb{R}^N$, a hypercube or in the case of periodic functions a circle.

There are many different notions of convergence for such a sequence of functions, we restrict our attention to the following four types:

Uniform convergence: The sequence $f_n(x)$ is said to converge uniformly to a function $f(x)$ defined on the same set Ω if for any positive ϵ there is an integer N such that

$$\left| f_n(x) - f(x) \right| < \epsilon \tag{3.6}$$

for all $n > N$ and for all $x \in \Omega$. In words: For high enough index n all functions f_n lie inside a band of width 2ϵ around f.

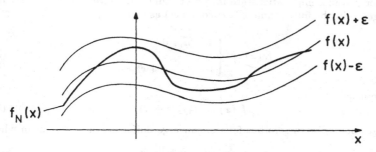

Fig. 3.2   Uniform convergence

Pointwise convergence: f_n converges pointwise to f if for every $x \in \Omega$ and any positive ϵ there is an integer N_x, which may depend on x, such that

$$\left| f_n(x) - f(x) \right| < \epsilon \tag{3.7}$$

for all $n > N_x$.

Convergence in the mean: f_n converges in the mean to f if for any positive ϵ there is an integer N such that

$$\int_\Omega \left| f_n(x) - f(x) \right|^2 dx < \epsilon \tag{3.8}$$

for all $n > N$. Here convergence is measured essentially by the volume between a function of the sequence and the limit function. Note that for periodic functions we integrate only over one period.

Weak convergence: f_n converges weakly to f if for any test function $T(x)$ and for any positive ϵ there is an integer N_T, which may depend on the test function, such that

$$\left| \int_\Omega (f_n(x) - f(x)) T(x) dx \right| < \epsilon \tag{3.9}$$

for all $n > N_T$. Note that in the case of periodic functions every differentiable function is already a test function because the circle is compact.

There are a few trivial implications: Uniform convergence implies pointwise and weak convergence. Convergence in the mean implies weak convergence due to the Cauchy-Schwarz inequality for integrals:

$$\left| \int_\Omega fT \right| \leq \sqrt{\int_\Omega |f|^2} \sqrt{\int_\Omega |T|^2}. \tag{3.10}$$

All other possible implications are incorrect: Uniform convergence does not imply convergence in the mean (unless Ω is compact) as shown by

$$f_n(x) := \begin{cases} 1/\sqrt{n}, & 0 < x < n \\ 0, & \text{elsewhere} \end{cases} \tag{3.11}$$

The sequence

$$f_n(x) := T_{1/2}(x - n) \tag{3.12}$$

converges pointwise to zero but fails to converge uniformly or in the mean whereas

$$f_n(x) := n^3 T_{1/2}(n^2(x - 1/n)) \tag{3.13}$$

converges pointwise to zero but diverges weakly. The sequence

$$f_n(x) := \sqrt{\frac{\sqrt{n}}{\pi} \frac{1}{1 + n^2 x^2}} \tag{3.14}$$

converges in the mean but not pointwise. Finally

$$f_n(x) := \sqrt{n} T_{1/2}(nx) \tag{3.15}$$

converges weakly to zero and fails to converge pointwise as well as in the mean. Note, however, that if a sequence of functions converges in two different modes to two continuous functions, then these two functions are equal.

In the calculation of the last section we interchanged limit and differentiation. For functions even the strongest form of convergence, the uniform one, is not enough to justify this operation. Consider for example the sequence

$$f_n(x) := \frac{1}{n} \sin n^2 x. \tag{3.16}$$

It converges uniformly to zero, but its derivatives

$$f_n'(x) = n \cos n^2 x \tag{3.17}$$

diverge.

Theorem: Let $f_n(x)$ be a sequence of once continuously differentiable functions. Suppose their derivatives $f'_n(x)$ converge uniformly towards a function $g(x)$. Furthermore suppose that at one point $x = a$, $f_n(a)$ converges. Then $f_n(x)$ converges uniformly to a function $f(x)$ and $f'(x) = g(x)$.

The proof relies on the interchange of uniform limits and integration.

Theorem: If a sequence of continuous functions $f_n(x)$ converges uniformly to the function $f(x)$, then $f(x)$ is continuous and $\int_a^x f_n(\xi)d\xi$ converges uniformly to $\int_a^x f(\xi)d\xi$.

This theorem does not apply to pointwise convergence as shown by the sequence

$$f_n(x) := x^n, \quad x \in [0, 1]. \tag{3.18}$$

Any deterrence at this point should be taken as motivation for weak convergence and distributions.

Exercises

1. In what sense do the following sequences converge?
   a) $f_n(x) = x/n$,   b) $f_n(x) = \sin nx$.
2. Verify one of the counter examples in this section.
3. Prove one of the theorems.

3.3 Real and complex harmonics

It is often convenient to work with complex exponentials rather than trigonometric functions, c.f. the remark preceding equation (1.43). Then the Fourier series of a function $f(t)$ with period T takes the complex form

$$f(t) = \lim_{n \to \infty} \sum_{j=-n}^{+n} c_j e^{-ij2\pi t/T}. \tag{3.19}$$

with the (complex) Fourier coefficients

$$c_j = \begin{cases} \frac{1}{2}(a_j + ib_j), & j > 0 \\ a_0, & j = 0 \\ \frac{1}{2}(a_{-j} - ib_{-j}), & j < 0. \end{cases} \tag{3.20}$$

Indeed using Euler's formula (1.37) we get

$$
\begin{aligned}
f(t) &= \sum_{j=-\infty}^{+\infty} c_j(\cos j\frac{2\pi}{T}t - \mathrm{i}\sin j\frac{2\pi}{T}t)\\
&= \sum_{j=1}^{\infty} c_{-j}(\cos j\frac{2\pi}{T}t + \mathrm{i}\sin j\frac{2\pi}{T}t)\\
&\quad + c_0 + \sum_{j=1}^{\infty} c_j(\cos j\frac{2\pi}{T}t - \mathrm{i}\sin j\frac{2\pi}{T}t)\\
&= c_0 + \sum_{j=1}^{\infty}(c_j + c_{-j})\cos j\frac{2\pi}{T}t - \sum_{j=1}^{\infty}\mathrm{i}(c_j - c_{-j})\sin j\frac{2\pi}{T}t,
\end{aligned} \tag{3.21}
$$

and

$$
a_0 = c_0, \tag{3.22}
$$
$$
a_j = c_j + c_{-j}, \tag{3.23}
$$
$$
b_j = -\mathrm{i}(c_j - c_{-j}). \tag{3.24}
$$

In particular if $f(t)$ is a real function, then the Fourier coefficients a_j and b_j are real and

$$
c_{-j} = \overline{c_j}. \tag{3.25}
$$

Exercise

1. What relation is satisfied by the complex Fourier coefficients of an even periodic function with complex values

$$
f(-t) = f(t) \quad ?
$$

3.4 A scalar product and the Fourier coefficients

In this section we calculate the Fourier coefficients of a periodic function. Here convergence in the mean is the technique of choice because its norm comes from a scalar product, defined for two complex valued, periodic functions f and g of period T:

$$
(f,g) := \int_0^T \bar{f}(t)g(t)dt. \tag{3.26}
$$

If the integrals exist, the scalar product is sesqui-linear, that means linear in its second component and anti-linear in the first:

$$
(f, g + \tilde{g}) = (f,g) + (f,\tilde{g}), \tag{3.27}
$$

$$(f, ag) = a(f, g), \quad a \in \mathbb{C}, \tag{3.28}$$

$$(f + \tilde{f}, g) = (f, g) + (\tilde{f}, g), \tag{3.29}$$

$$(af, g) = \bar{a}(f, g). \tag{3.30}$$

According to the theorem that whenever several conventions are possible all of them are represented in the literature, many authors define the scalar product with the complex conjugation on the second function g and thus anti-linearity in the second component. In any case

$$(f, g) = \overline{(g, f)} \tag{3.31}$$

and the scalar product is positive

$$(f, f) \geq 0. \tag{3.32}$$

Its norm

$$|f| := \sqrt{(f, f)} \tag{3.33}$$

is the one that measures convergence in the mean.

Let us introduce a short-hand for the complex harmonics,

$$e_j(t) := \frac{1}{\sqrt{T}} e^{-ij2\pi t/T}, \quad j \in \mathbb{Z}. \tag{3.34}$$

The normalization $1/\sqrt{T}$ is chosen to make the $e_j(t)$ orthonormal with respect to the scalar product:

$$(e_j, e_k) = \delta_{jk}. \tag{3.35}$$

Indeed

$$|e_j|^2 = \int_0^T \frac{1}{\sqrt{T}} e^{ij2\pi t/T} \frac{1}{\sqrt{T}} e^{-ij2\pi t/T} dt = \frac{1}{T} \int_0^T dt = 1 \tag{3.36}$$

and for $j \neq k$

$$(e_j, e_k) = \int_0^T \frac{1}{\sqrt{T}} e^{ij2\pi t/T} \frac{1}{\sqrt{T}} e^{-ik2\pi t/T} dt =$$
$$= \frac{1}{T} \int_0^T e^{-i(k-j)2\pi t/T} dt = \frac{1}{-i(k-j)2\pi} e^{-i(k-j)2\pi t/T} \Big|_0^T = 0. \tag{3.37}$$

For a given periodic function $f(t)$ consider the n-th approximation of the Fourier series and let us see how well it approximates f in the mean:

$$\left| f - \sum_{j=-n}^{n} c_j \sqrt{T} e_j \right|^2$$

$$= (f - \sqrt{T} \sum_{j=-n}^{n} c_j e_j, f - \sqrt{T} \sum_{k=-n}^{n} c_k e_k)$$

$$= (f,f) - \sqrt{T} \sum_k c_k(f,e_k) - \sqrt{T} \sum_j \bar{c}_j(e_j,f) + T \sum_{j,k} \overline{c_j} c_k(e_j,e_k) \qquad (3.38)$$

$$= (f,f) - \sqrt{T} \sum_j c_j \overline{(e_j,f)} - \sqrt{T} \sum_j \overline{c_j}(e_j,f) + T \sum_j \overline{c_j} c_j$$

$$= (f,f) - \sum_j |(f,e_j)|^2 + \sum_j |\sqrt{T} c_j - (e_j,f)|^2$$

where we have used sesqui-linearity of the scalar product and orthonormality of the harmonics. We must choose the Fourier coefficients such that this norm is as small as possible. Therefore

$$c_j = \frac{1}{\sqrt{T}}(e_j,f) = \frac{1}{T} \int_0^T f(t) e^{\mathrm{i} j 2\pi t/T} dt \qquad (3.39)$$

and

$$\left| f - \sum_{j=-n}^{n} c_j e^{-\mathrm{i} j 2\pi t/T} \right|^2 = \int_0^T |f(t)|^2 dt - T \sum_{j=-n}^{n} |c_j|^2 \geq 0 \qquad (3.40)$$

$$\text{"Bessel's inequality"}.$$

Furthermore we have
Parseval's theorem: If the periodic function $f(t)$ is square integrable,

$$\int_0^T |f(t)|^2 dt < \infty,$$

its Fourier series with coefficients given by equation (3.39) converges in the mean towards f and

$$\int_0^T |f(t)|^2 dt = T \sum_{j=-\infty}^{\infty} |c_j|^2. \qquad (3.41)$$

In other words, the harmonics $e_j(t)$ form an orthonormal basis in the space of square integrable periodic functions.

Example: Consider the periodic function of period 1 defined by $f(t) := t - \frac{1}{2}$ in the half open interval $t \in [0,1)$. It is square-integrable:

$$\int_0^1 \left| t - \frac{1}{2} \right|^2 dt = \frac{1}{12}. \qquad (3.42)$$

Its Fourier coefficients are calculated using an integration by parts for $j \neq 0$:

$$c_j = \int_0^1 (t - \frac{1}{2}) e^{ij2\pi t} dt = t \frac{1}{2\pi i j} e^{2\pi i j t} \Big|_0^1 - \int_0^1 \frac{e^{2\pi i j t}}{2\pi i j} - \frac{1}{2} \int_0^1 e^{2\pi i j t} dt$$

$$= \int_0^1 \frac{e^{2\pi i j t}}{2\pi i j} - \frac{1}{2} \int_0^1 e^{2\pi i j t} dt = \frac{1}{2\pi i j} e^{2\pi i j} = \frac{1}{2\pi i j} \qquad (3.43)$$

and

$$c_0 = \int_0^1 (t - \frac{1}{2}) dt = 0. \qquad (3.44)$$

Therefore

$$f(t) = \frac{1}{2\pi i} \sum_{\substack{j=-\infty \\ j \neq 0}}^{+\infty} \frac{1}{j} e^{-2\pi i j t} = \frac{-1}{\pi} \sum_{j=1}^{\infty} \frac{1}{j} \sin 2\pi j t. \qquad (3.45)$$

where the limit is understood in the mean.

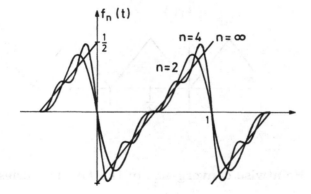

Fig. 3.3 The harmonic saw and its Fourier series $f_n(t) := \frac{-1}{\pi} \sum_{j=1}^{n} \frac{1}{j} \sin 2\pi j t$

The Fourier coefficients form a harmonic sequence and the function $f(t)$ is also referred to as harmonic saw. Made audible it sounds indeed harmonic and is used in synthezisers. The coefficients a_j all vanish because $f(t)$ is an odd function. Parseval's equation reads

$$\sum_{j=1}^{\infty} \frac{1}{j^2} = \frac{\pi^2}{6}, \qquad (3.46)$$

an interesting result in itself.

We close this section with an immediate consequence of Parseval's theorem:

Corollary: If $f(t)$ and $g(t)$ are two periodic square integrable functions with Fourier coefficients c_j and k_j then

$$\int_0^T \bar{f}(t)g(t)dt = T \sum_{j=-\infty}^{+\infty} \overline{c_j}k_j. \tag{3.47}$$

The proof uses the Cauchy-Schwarz inequality for sums and for integrals.

Exercises

1. Calculate the real Fourier series of $\sin^2$.
2. Calculate the complex and real Fourier coefficients of
   a)
   $$f(t) := \begin{cases} -1, & \ell < t < \ell + \frac{1}{2} \\ +1, & \ell + \frac{1}{2} < t < \ell + 1 \end{cases} \quad \ell \in \mathbb{Z},$$
   b)

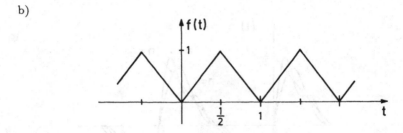

3. Prove the corollary.

3.5 Pointwise convergence of the Fourier series

There is no reason why the Fourier series in the last example should converge pointwise, or if it does that its limit is the value $f(t)$. Indeed for $t = 0$ the Fourier series (3.45) converges to zero while $f(0) = -\frac{1}{2}$. This is no surprise, just as for weak convergence the limit of convergence in the mean cannot be localized in the sense that redefining a function in a given point t_0 does not affect its Fourier coefficients, nor convergence in the mean. Our main interest is in distributions, only for completeness and later proofs do we cite two theorems about pointwise and uniform convergence of the Fourier series.

Theorem: If a periodic function $f(t)$ is piecewise differentiable, then its Fourier series converges pointwise to the mid-values

$$m(t) := \lim_{h \to 0+} \frac{1}{2}[f(t+h) + f(t-h)]. \tag{3.48}$$

The Fourier series of the harmonic saw at the jump $t = 0$ indeed converges to the mid-value $m(0) = 0$. At $t = \frac{1}{4}$ where the saw is continuous the function coincides with its mid-value and the theorem yields the so-called Leibniz series

$$\sum_{j=1}^{\infty} \frac{1}{2j-1} (-1)^{j+1} = \frac{\pi}{4}. \tag{3.49}$$

Theorem: If a periodic function $f(t)$ is once continuously differentiable, then its Fourier series converges uniformly and absolutely. Its limit is $f(t)$.

Exercises:

1. Examine the pointwise convergence of the Fourier series for the functions in the last exercise 2.
2. Calculate the Fourier series of the periodic function with period 1 defined by

$$f(t) = \begin{cases} -4t^3 + 3t^2, & 0 \le t \le \frac{1}{2} \\ 4t^3 - 9t^2 + 6t - 1, & \frac{1}{2} \le t \le 1 \end{cases}$$

What are its modes of convergence?
3. Is there a differentiable function with Fourier coefficients $c_j = 1/(j^2 + 1)$?
4. Prove the second theorem.
   Hint: Integrate by parts the rhs of equation (3.39) and use the Cauchy-Schwarz inequality for sums together with Bessel's inequality for $f'(t)$.

3.6 Fourier series of periodic distributions

A distribution in one real variable $D(t)$ is called periodic with period T if

$$D(t + T) = D(t). \tag{3.50}$$

For example consider

$$D(t) := \sum_{j=-\infty}^{+\infty} \delta_j(t). \tag{3.51}$$

It is a distribution as convergent sequence of distributions and it is periodic with period one. As in the case of functions, periodic distributions can equivalently be viewed as distributions on the circle with circumference T. In this setting the distribution of the example can be represented by the weakly convergent sequence of differentiable functions on the circle of unit circumference:

$$f_n(t) := \frac{T_{1/(3n)}(t)}{\int_{-1/2}^{1/2} T_{1/(3n)}(\tau) d\tau}, \quad n = 1, 2, 3, \ldots \tag{3.52}$$

In chapter 2 we defined an operation for a distribution by first considering this operation for functions representing the distribution and then taking the limit. In this spirit we define the Fourier coefficients of a periodic distribution $D(t)$ by considering the Fourier coefficient of periodic functions $f_n(t)$ which represent $D(t)$:

$$c_j^n := \frac{1}{T} \int_0^T f_n(t) e^{ij2\pi t/T} dt. \tag{3.53}$$

We have to show that for every j the limit as n tends to infinity exists and does not depend on the particular choice of the weakly convergent sequence $f_n(t)$ used to represent the distribution $D(t)$. To this end it suffices to remember that the harmonics are test functions on the circle. Therefore we can define the Fourier coefficients of a periodic distribution $D(t)$ by:

$$c_j = \lim_{n \to \infty} c_j^n = \lim_{n \to \infty} \frac{1}{T} \int_0^T f_n(t) e^{ij2\pi t/T} dt$$
$$= \frac{1}{T} \int_0^T D(t) e^{ij2\pi t/T} dt = \frac{1}{\sqrt{T}} D(e_{-j}), \quad j \in \mathbb{Z}. \tag{3.54}$$

For example consider the periodic δ-distribution with period 1, equation (3.51). Its Fourier coefficients are

$$c_j = \int_0^1 \delta(t) e^{ij2\pi t/T} dt = 1. \tag{3.55}$$

Let us show that its Fourier series converges (in the sense of distributions) to the δ-distribution, i.e. we must show that for any test function $T(t)$ on the circle

$$\lim_{n \to \infty} \int_0^1 \sum_{j=-n}^n e^{-2\pi i jt} T(t) dt = T(0). \tag{3.56}$$

To this end we use the uniform convergence of the Fourier series of the test function

$$T(x) = \lim_{m \to \infty} \sum_{k=-m}^m d_k e^{-2\pi i kx} \tag{3.57}$$

and interchange uniform limit and integration:

$$\lim_{n\to\infty} \int_0^1 \sum_{j=-n}^{n} e^{-2\pi ijt} T(t)dt$$

$$= \lim_{n\to\infty} \sum_{j=-n}^{n} \int_0^1 e^{-2\pi ijt} \lim_{m\to\infty} \sum_{k=-m}^{m} d_k e^{-2\pi ikt} dt$$

$$= \lim_{n\to\infty} \lim_{m\to\infty} \sum_{j,k} d_k \int_0^1 e^{-2\pi ijt} e^{-2\pi ikt} dt \qquad (3.58)$$

$$= \lim_{n\to\infty} \lim_{m\to\infty} \sum_{j,k} d_k \delta_{-j,k} = \lim_{m\to\infty} \sum_{k=-m}^{m} d_k$$

$$= \sum_{k=-\infty}^{\infty} d_k e^{-2\pi ik0} = T(0).$$

Therefore

$$\sum_{j=-\infty}^{\infty} e^{-2\pi ijt} = \delta(t). \qquad (3.59)$$

An alternative proof uses the Fourier series of the harmonic saw on the circle parametrized by $t \in [-\frac{1}{2}, \frac{1}{2}]$:

$$\frac{1}{2\pi i} \sum_{\substack{j=-\infty \\ j\neq 0}}^{\infty} \frac{1}{j} e^{-2\pi ijt} = \begin{cases} t + \frac{1}{2}, & -\frac{1}{2} \le t \le 0 \\ t - \frac{1}{2}, & 0 \le t \le \frac{1}{2}. \end{cases} \qquad (3.60)$$

This limit is also valid in the sense of distributions and as such may be interchanged with differentiation. Taking one derivative on both sides, we obtain

$$\frac{1}{2\pi i} \sum_{\substack{j=-\infty \\ j\neq 0}}^{\infty} \frac{-2\pi ij}{j} e^{-2\pi ijt} = 1 - \delta(t) \qquad (3.61)$$

or

$$\delta(t) = \sum_{j=-\infty}^{\infty} e^{-2\pi ijt} = 1 + 2 \sum_{j=1}^{\infty} \cos 2\pi jt. \qquad (3.62)$$

As curiosity we note that Euler assigned the value zero to this series. Differentiating once more we obtain the Fourier coefficients of $\delta'(t)$:

$$c_j = -2\pi ij. \qquad (3.63)$$

For any periodic distribution we have the very convenient

Theorem: The Fourier series of a periodic distribution $D(t)$ converges to $D(t)$:

$$\sum_{j=-\infty}^{\infty} c_j e^{-ij2\pi t/T} = D(t) \qquad (3.64)$$

with

$$c_j := \frac{1}{T} \int_0^T D(t) e^{ij2\pi t/T} dt. \qquad (3.65)$$

By equation (2.60) this Fourier series may be differentiated term by term and differentiation with respect to t amounts to multiplication with $-ij2\pi/T$ in "frequency space".

The proof of the general theorem again uses the uniform convergence of the Fourier series of any test function $T(t)$, equation (3.57). As $T(t)$ is infinitely many times differentiable, its Fourier coefficients d_k are of fast decrease. This means that d_k multiplied by any power of k still tends to zero with increasing k and

$$\int D(t)\bar{T}(t)dt = T \sum_{j=-\infty}^{+\infty} c_j \overline{d_j} \qquad (3.66)$$

converges for c_j a polynomial in j. Note the similarity between this equation and the corollary to Parseval's theorem, equation (3.47), explaining why many authors use the scalar product notation for the evaluation of a distribution on a test function:

$$\int D(t)\bar{T}(t)dt =: \overline{(D,T)}. \qquad (3.67)$$

We shall not follow this convention, but we shall come back to this interplay between weak convergence and convergence in the mean in chapter 6. Before passing to applications let us summarize this chapter in the following table:

Note that the first and last line are dual to each other in the sense of equation (3.66) while the fourth line is self-dual.

Exercises

1. If c_j are the Fourier coefficients of the distribution $D(t)$ with period T, show that the Fourier coefficient of $D(t-a)$ are $e^{ij2\pi a/T} c_j$.
2. What relations characterize the Fourier coefficients of
   a) a real periodic distribution,
   b) an even periodic distribution,
   c) a purely imaginary, odd distribution?
3. Recalculate the Fourier coefficients of the periodic function in ex. 2, section 3.5, by differentiating it four times before calculating Fourier coefficients.
4. Calculate the Fourier coefficients of the function defined by $f(t) = t^2$, $-1 \le t \le 1$ with period 2 as well as of all its derivatives.

property of periodic function	type of convergence of its Fourier series	property of its Fourier coefficients	typical example with period 1
differentiable (test function)	uniform	fast decrease	$f(t) = \dfrac{1-(1/13)\cos 2\pi t}{1-(2/13)\cos 2\pi t+(1/13)^2}$ $c_j = \frac12\left(\frac{1}{13}\right)^{|j|},\quad j \neq 0$ $c_0 = 1$
once continuously differentiable	uniform	$\sum_{j=-\infty}^{+\infty}|c_j| < \infty$ only a necessary condition	$f(t) = \begin{cases} -4t^3 + 3t^2, & 0 \le t \le \frac12 \\ 4t^3 - 9t^2 + 6t - 1, & \frac12 \le t \le 1 \end{cases}$ $c_j = \frac{3}{\pi^4}\frac{(-1)^j-1}{j^4}\quad j \neq 0$ $c_0 = \frac18$
piecewise differentiable	pointwise to the midvalue	$\sum_{j=-\infty}^{+\infty} c_j < \infty$ only a necessary condition	$f(t) = t - \frac12,\ 0 < t < 1$ $c_j = \frac{1}{2\pi i j},\quad j \neq 0$ $c_0 = 0$
$\int_0^T |f(t)|^2\,dt < \infty$	in the mean	$\sum_{j=-\infty}^{+\infty}|c_j|^2 < \infty$	$f(t) = \frac12 \ln \frac{1}{2(1-\cos 2\pi t)}$ $c_j = \frac{1}{2|j|},\quad j \neq 0$ $c_0 = 0$
distribution	as distribution	polynomial increase or slower	$\delta''(t)$ $c_j = (2\pi i j)^2$

Tab. 3.1   Fourier series

5. Show that no distribution can have Fourier coefficients $c_j = |j|!$
6. Indicate a periodic function which is not continuously differentiable and whose Fourier coefficients satisfy $\sum_{j=-\infty}^{\infty} |c_j| < \infty$.

3.7 Forced oscillations

Let us apply a periodic force to the damped harmonic oscillator. Now the source function extends infinitely into the past and we cannot use the magic formula (2.90). Instead the Fourier series will solve the problem. As most applications of forced oscillations come from electricity we write the harmonic oscillator here in its electric form: Consider an electric circuit consisting of a coil with self-inductance L, a capacitor with capacity C and a resistor with resistance R connected in series to a generator of periodic electromotive force $V(t)$.

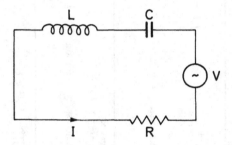

Fig. 3.4   Electric circuit

The resultant electric current $I(t)$ in the circuit is given by the inhomogeneous differential equation of the harmonic oscillator:

$$L\frac{d^2}{dt^2}I + R\frac{d}{dt}I + \frac{I}{C} = \frac{d}{dt}V. \tag{3.68}$$

If T is the period of the electromotive force we define the frequency

$$\nu := \frac{1}{T} \tag{3.69}$$

and the angular frequency

$$\omega := \frac{2\pi}{T}. \tag{3.70}$$

Let us develop $V(t)$ in its Fourier series.

$$V(t) = \sum_{j=-\infty}^{\infty} V_j e^{-ij\omega t}, \tag{3.71}$$

$$V_j = \frac{1}{T} \int_0^T V(t)e^{ij\omega t}. \tag{3.72}$$

Knowing from our childhood that the response $I(t)$ has the same frequency as the source, we try

$$I(t) = \sum_{j=-\infty}^{+\infty} I_j e^{-ij\omega t}. \tag{3.73}$$

Plugging this ansatz into our differential equation, we obtain in the sense of distributions:

$$(-j^2\omega^2 L - ij\omega R - \frac{1}{C})I_j = -ij\omega V_j. \tag{3.74}$$

or

$$I_j = \frac{1}{R - i(j\omega L - \frac{1}{j\omega C})}V_j = \frac{V_j}{z_j}. \tag{3.75}$$

The complex numbers z_j determine the current. The absolute value of z_j is called impedance and fixes the amplitude of I_j, its polar angle is the phase difference between the j-th components of current and voltage. Note that the impedance is minimal $z_j = R$ if the angular frequency of the source coincides with the angular frequency of the free oscillator

$$\omega_0 := \sqrt{\frac{1}{LC}}. \tag{3.76}$$

Therefore our circuit filters out frequencies close to the natural frequency $\omega_0/(2\pi)$.

Exercise

1. Find all periodic sources $V(t)$ that produce a current of the same shape:

$$I(t) = bV(t-a) \quad a, b \in \mathbb{R}.$$

3.8 The string

In exercise 4 of section 2.6 we solved the equation of the static string under an external force. Let us now introduce dynamics. We denote by L the length of the string, by T its tension, $\rho(x)$ is its linear mass density, $F(t,x)$ an external transverse force density, and $u(t,x)$ its transverse displacement.

In the approximation of small forces, $F \ll T/L$, and neglecting friction, the displacement satisfies the one-dimensional inhomogeneous wave equation

$$\frac{1}{c^2}\frac{\partial^2 u}{\partial t^2} - \frac{\partial^2 u}{\partial x^2} = -\frac{F}{T} \tag{3.77}$$

72

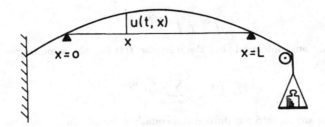

Fig. 3.5   The string

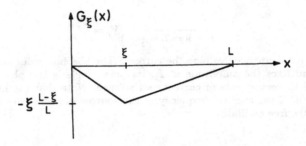

Fig. 3.6   The static Green function

where

$$c = \sqrt{\frac{T}{\rho}} \qquad (3.78)$$

is the propagation speed of a transverse wave in the string. The wave equation is a second-order linear partial differential equation. It admits a unique solution subject to the boundary condition

$$u(t,0) = u(t,L) \equiv 0 \qquad (3.79)$$

for all times and with specified initial condition

$$u(0,x) = g(x), \qquad (3.80)$$

$$\frac{\partial u}{\partial t}(0,x) = h(x). \qquad (3.81)$$

Picking a guitar string, for instance, is described by f proportional to the static Green function

$$G_\xi(x) = \begin{cases} -x(L-\xi)/L, & 0 \le x \le \xi \\ -\xi(L-x)/L, & \xi \le x \le L \end{cases} \qquad (3.82)$$

and vanishing h, while for a piano string g is zero.

From now on we suppose c constant and vanishing external force $F \equiv 0$. We solve the wave equation by the ansatz

$$u(t, x) = v(t)w(x), \tag{3.83}$$

called separation of variables. Inserting the ansatz into the wave equation, we obtain

$$\frac{1}{c^2}\frac{1}{v}\frac{d^2v}{dt^2} = \frac{1}{w}\frac{d^2w}{dx^2}. \tag{3.84}$$

The lhs only depends on t, the rhs on x only. Therefore they must both be equal to a real constant λ. The separation of variables has reduced the wave equation to two harmonic oscillator equations one in t, one in x.

$$\frac{d^2v}{dt^2} - \frac{\lambda}{c^2}v = 0, \tag{3.85}$$

$$\frac{d^2w}{dx^2} - \lambda w = 0. \tag{3.86}$$

The second is subject to the boundary condition

$$w(0) = w(L) = 0. \tag{3.87}$$

For positive λ the most general solution of (3.86)

$$w(x) = Ae^{\sqrt{\lambda}x} + Be^{-\sqrt{\lambda}x}, \qquad A, B \in \mathbb{R} \tag{3.88}$$

cannot satisfy the boundary condition unless $A = B = 0$. If λ vanishes, we have the solution

$$w(x) = Ax + B \tag{3.89}$$

again incompatible with the boundary condition. Only for negative λ where

$$w(x) = A\cos\sqrt{-\lambda}x + B\sin\sqrt{-\lambda}x \tag{3.90}$$

can we satisfy the boundary condition by putting $A = 0$ and

$$\sqrt{-\lambda} = j\frac{\pi}{L}, \quad j = 1, 2, 3, \dots \tag{3.91}$$

Therefore we have a sequence of solutions of equation (3.86)

$$w_j(x) = B_j \sin j\frac{\pi}{L}x. \tag{3.92}$$

The corresponding $v(t)$ must solve

$$\frac{d^2v_j}{dt^2} + j^2\frac{c^2\pi^2}{L^2}v_j = 0. \tag{3.93}$$

Therefore

$$v_j(t) = C_j \cos j\frac{\pi c}{L}t + D_j \sin j\frac{\pi c}{L}t \qquad (3.94)$$

with real constants C_j and D_j. The wave equation being linear we may superpose these solutions

$$u(t, x) = \sum_{j=1}^{\infty} (E_j \cos j\omega t + F_j \sin j\omega t) \sin j\frac{\pi}{L}x \qquad (3.95)$$

with angular frequency

$$\omega := \frac{\pi c}{L}. \qquad (3.96)$$

We now have to answer the question: was our ansatz sufficiently general to satisfy the initial conditions (3.80) and (3.81) and how do they determine the constants E_j and F_j:

$$u(0, x) = \sum_{j=1}^{\infty} E_j \sin j\frac{2\pi}{2L}x = g(x), \qquad (3.97)$$

$$\frac{\partial u}{\partial t}(0, x) = \sum_{j=1}^{\infty} j\omega F_j \sin j\frac{2\pi}{2L}x = h(x)? \qquad (3.98)$$

Note that in the last equation we have already interchanged differentiation and limit, which is justified in the sense of distributions. The above question can now be answered using a trick: Define two periodic functions of period $2L$ by

$$\tilde{g}(x) = \begin{cases} g(x), & 0 \le x \le L \\ -g(2L - x), & L \le x \le 2L. \end{cases} \qquad (3.99)$$

and analoguously for h.

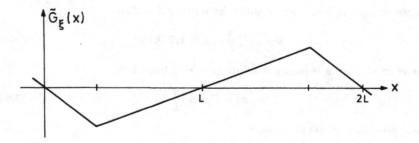

Fig. 3.7   The periodic extension

Note that if g is differentiable, so is $\tilde{g}$. From the symmetry of $\tilde{g}$ we have that its Fourier coefficients a_j all vanish.

$$\tilde{g}(x) = \sum_{j=1}^{\infty} b_j \sin j \frac{2\pi}{2L} x. \tag{3.100}$$

Therefore

$$E_j = b_j = \frac{1}{L} \int_0^{2L} \tilde{g}(x) \sin j \frac{2\pi}{2L} x \, dx = \frac{2}{L} \int_0^L g(x) \sin j \frac{\pi}{L} x \, dx. \tag{3.101}$$

and similarly

$$F_j = \frac{2}{j\omega L} \int_0^L h(x) \sin j \frac{\pi}{L} x \, dx. \tag{3.102}$$

For the guitar string all F_j's vanish, E_1 is the amplitude of the keynote, E_2 the amplitude of the first octave and so on. It is known that the timbre of a guitar depends on how close to the bridge the string is picked. Indeed if we pick the string at ξ as in Fig. 3.6, then the amplitudes are given by

$$E_j = \frac{-2L}{j^2 \pi^2} \sin j \pi \frac{\xi}{L} \tag{3.103}$$

in particular picking in the middle, $\xi = \frac{L}{2}$, produces no odd octaves $E_2 = 0$, $E_4 = 0, \dots$.

Exercises

1. Verify equation (3.102).
2. Verify equation (3.103).

4

Fourier transforms

4.1 Heuristics

We try to generalize the Fourier series to functions which are not necessarily periodic by letting the period tend to infinity after a clever redefinition of variables. Let $f(x) : \mathbb{R} \to \mathbb{C}$ be a function with period L. Its Fourier series is

$$f(x) = \sum_{j=-\infty}^{+\infty} c_j e^{-ij2\pi x/L} \tag{4.1}$$

with Fourier coefficients

$$c_j = 1/L \int_{-L/2}^{+L/2} f(x) e^{ij2\pi x/L} dx. \tag{4.2}$$

We define the "wave number"

$$k := j\frac{2\pi}{L}. \tag{4.3}$$

(If the independent variable of f is t instead of x, it is tradition to denote the period by T and to use the letter ω instead of k

$$\omega := j\frac{2\pi}{T} \tag{4.4}$$

then called angular frequency.) We also redefine the dependent variable c as a function of j

$$c_j =: \frac{\sqrt{2\pi}}{L} \hat{f}\left(\frac{2\pi}{L}j\right) \tag{4.5}$$

or

$$\hat{f}(k) = \frac{L}{\sqrt{2\pi}} c_{Lk/(2\pi)}. \tag{4.6}$$

Now the Fourier series

$$f(x) = \frac{1}{\sqrt{2\pi}} \sum_{j=-\infty}^{+\infty} \hat{f}(j\frac{2\pi}{L}) e^{-ij2\pi x/L}\frac{2\pi}{L} = \frac{1}{\sqrt{2\pi}} \sum \hat{f}(k) e^{-ikx} \Delta k \tag{4.7}$$

takes the form of a Riemann sum and as L tends to infinity we obtain the Fourier integral

$$f(x) = \frac{1}{\sqrt{2\pi}} \int_{-\infty}^{+\infty} \hat{f}(k)e^{-ikx}dk. \tag{4.8}$$

Simultaneously the Fourier coefficients in the new variables become

$$\hat{f}(k) = \frac{1}{\sqrt{2\pi}} \int_{-L/2}^{+L/2} f(x)e^{ikx}dx. \tag{4.9}$$

and in the limit of large L

$$\hat{f}(k) = \frac{1}{\sqrt{2\pi}} \int_{-\infty}^{+\infty} f(x)e^{ikx}dx. \tag{4.10}$$

In this limit the discrete independent variable j is replaced by the continuous variable k and we shall see that if $f(x)$ is a well-behaved function, then $\hat{f}(k)$ will also be a well behaved function.

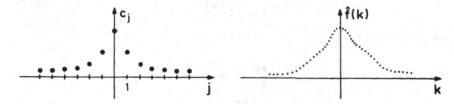

Fig. 4.1   As the period goes to infinity

4.2 The Fourier transform of functions

In this section we explain the adjective "well-behaved".

Definition: A function $f : \mathbb{R} \rightarrow \mathbb{C}$ is said to be of fast decrease (or Schwartz function) if f is differentiable, if the product of f by any polynomial is bounded and if the product of any derivative of f by any polynomial is bounded.

Any test function is of fast decrease as well as the Gaussian

$$f(x) := e^{-x^2/a^2}, \quad a > 0. \tag{4.11}$$

If f is of fast decrease, then so is any derivative of f and the product of f by any polynomial.

78

Definition: We define the Fourier transform of a function of fast decrease $f(x)$ by the convergent integral

$$\hat{f}(k) := \frac{1}{\sqrt{2\pi}} \int_{-\infty}^{+\infty} f(x) e^{ikx} dx. \qquad (4.12)$$

We should note that other conventions for the Fourier transform are common with opposite sign in the exponent and/or beauty factors $\sqrt{2\pi}$ in other places.

bf Example:
$$f(x) = e^{-x^2/a^2}, \quad a > 0. \qquad (4.13)$$

We have to evaluate the integral

$$\hat{f}(k) := \frac{1}{\sqrt{2\pi}} \int_{-\infty}^{+\infty} e^{-x^2/a^2 - ikx} dx. \qquad (4.14)$$

To this end we "complete the square" in the exponent

$$\frac{x^2}{a^2} - ikx = \left(\frac{x}{a} - \frac{i}{2}ka\right)^2 + \left(\frac{1}{2}ka\right)^2 \qquad (4.15)$$

and apply Cauchy's theorem to the closed curve C of Fig. 4.2.

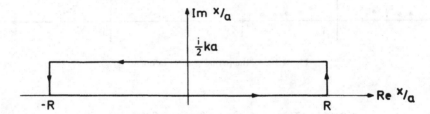

Fig. 4.2 Completing squares

$$\begin{aligned}
\hat{f}(k) &= \frac{1}{\sqrt{2\pi}} e^{-(ka/2)^2} \int_{-\infty}^{+\infty} e^{-(x/a - ika/2)^2} dx \\
&= \frac{1}{\sqrt{2\pi}} e^{-(ka/2)^2} \int_{-\infty}^{+\infty} e^{-t^2} a\, dt = \frac{a}{\sqrt{2}} e^{-a^2 k^2/4}.
\end{aligned} \qquad (4.16)$$

Therefore the Fourier transform of a Gaussian is a Gaussian. Both Gaussians have the same "norm":

$$\int_{-\infty}^{+\infty} |f(x)|^2 dx = \sqrt{\frac{\pi}{2}} a = \int_{-\infty}^{+\infty} |\hat{f}(k)|^2 dk, \qquad (4.17)$$

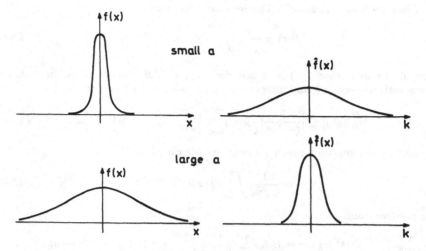

Fig. 4.3   The uncertainty principle

an equation which we call Parseval's equation again. The width of the first Gaussian is

$$\sqrt{\frac{\int_{-\infty}^{+\infty} x^2 |f(x)|^2 dx}{\int_{-\infty}^{+\infty} |f(x)|^2 dx}} = \frac{a}{2} \tag{4.18}$$

while the second has width $1/a$. In other words, the more $f(x)$ is peaked, the flatter is its Fourier transform.

This property is of general nature and will be proved in chapter 6. It expresses the uncertainty principle in quantum mechanics. Finally let us remark that our example satisfies

$$\hat{\hat{f}}(x) = f(-x), \tag{4.19}$$

or equivalently the Fourier inversion formula

$$f(x) = \frac{1}{\sqrt{2\pi}} \int_{-\infty}^{+\infty} \hat{f}(k) e^{-ikx} dk. \tag{4.20}$$

Parseval's equation and Fourier's inversion formula hold for any function of fast decrease:

Fourier's inversion theorem: Let f be a function of fast decrease. Then its Fourier transform

$$\hat{f}(k) := \frac{1}{\sqrt{2\pi}} \int_{-\infty}^{+\infty} f(x) e^{ikx} dx \tag{4.21}$$

is of fast decrease and satisfies Fourier's inversion formula

$$f(x) = \frac{1}{\sqrt{2\pi}} \int_{-\infty}^{+\infty} \hat{f}(k) e^{-ikx} dx. \tag{4.22}$$

Proof: To show that $\hat{f}(k)$ is of fast decrease, we differentiate equation (4.21) q times with respect to k and integrate by parts p times with respect to x

$$\left| \frac{d^q \hat{f}}{dk^q}(k) \right| \le \left| \frac{1}{(-ik)^p} \right| \frac{1}{\sqrt{2\pi}} \int_{-\infty}^{+\infty} \left| \frac{d^p}{dx^p} \left[(ix)^q f(x) \right] \right| \left| e^{ikx} \right| dx. \tag{4.23}$$

To prove the inversion formula we need two identities:

$$f(x) = \frac{1}{2\sqrt{\pi\epsilon}} \int_{-\infty}^{+\infty} f(x) e^{-(x-\xi)^2/(4\epsilon)} d\xi \tag{4.24}$$

for positive ϵ and

$$\frac{1}{\sqrt{2\pi}} \int_{-\infty}^{+\infty} e^{-\epsilon k^2} \hat{f}(k) e^{-ikx} dk = \frac{1}{2\pi} \int_{-\infty}^{+\infty} \left[\int_{-\infty}^{+\infty} f(\xi) e^{ik\xi} d\xi \right] e^{-\epsilon k^2 - ikx} dk$$

$$= \frac{1}{2\pi} \int_{-\infty}^{+\infty} f(\xi) \left[\int_{-\infty}^{+\infty} e^{-\epsilon k^2 - ik(x-\xi)} dk \right] d\xi. \tag{4.25}$$

The interchange of integration is justified by Fubini's theorem because the double integral is absolutely convergent. We have already evaluated the inner integral:

$$\int_{-\infty}^{+\infty} e^{-\epsilon k^2 - ik(x-\xi)} dk = \sqrt{\frac{\pi}{\epsilon}} e^{-(x-\xi)^2/(4\epsilon)}. \tag{4.26}$$

Hence we arrive at the second identity

$$\frac{1}{\sqrt{2\pi}} \int_{-\infty}^{+\infty} e^{-\epsilon k^2} \hat{f}(k) e^{-ikx} dk = \frac{1}{2\sqrt{\pi\epsilon}} \int_{-\infty}^{+\infty} f(\xi) e^{-(x-\xi)^2/(4\epsilon)} d\xi. \tag{4.27}$$

Finally the difference between $f(x)$ and the regularized Fourier transform of $\hat{f}(k)$ becomes

$$\left| \frac{1}{\sqrt{2\pi}} \int_{-\infty}^{+\infty} e^{-\epsilon k^2} \hat{f}(k) e^{-ikx} dk - f(x) \right|$$

$$\le \frac{1}{2\sqrt{\pi\epsilon}} \int_{-\infty}^{+\infty} \left| f(\xi) - f(x) \right| e^{-(x-\xi)^2/(4\epsilon)} d\xi$$

$$\le \max_{\xi \in \mathbb{R}} \left| f'(\xi) \right| \frac{1}{2\sqrt{\pi\epsilon}} \int_{-\infty}^{+\infty} |\xi - x| e^{-(x-\xi)^2/(4\epsilon)} d\xi \tag{4.28}$$

$$\le 2\sqrt{\frac{\epsilon}{\pi}} \max_{\xi \in \mathbb{R}} \left| f'(\xi) \right| \xrightarrow{\epsilon \to 0} 0.$$

Parseval's theorem: If f and g are functions of fast decrease with Fourier transforms $\hat{f}$ and $\hat{g}$, then

$$\int_{-\infty}^{+\infty} \bar{f}(x)g(x)dx = \int_{-\infty}^{+\infty} \bar{\hat{f}}(k)\hat{g}(k)dk. \tag{4.29}$$

Proof: Both sides are equal to the absolutely convergent double integral

$$\frac{1}{\sqrt{2\pi}} \int_{-\infty}^{+\infty} \int_{-\infty}^{+\infty} \bar{f}(x)\hat{g}(k)e^{-ikx}dxdk. \tag{4.30}$$

As for Fourier series, differentiation with respect to x amounts to multiplication by $-ik$ of the Fourier transform and translating f by a yields multiplication of $\hat{f}$ by e^{ika}.

Theorem: a) Let f be of fast decrease. Then

$$\widehat{\frac{df}{dx}}(k) = -ik\widehat{f}(k), \tag{4.31}$$

$$(\widehat{ixf})(k) = \frac{d\hat{f}}{dk}(k). \tag{4.32}$$

b) If $f(x) = g(x - a)$, then

$$\hat{f}(k) = e^{ika}\hat{g}(k). \tag{4.33}$$

c) If f is even (odd), then so is its Fourier transform.

d) If f and g are of fast decrease, then their convolution product

$$(f * g)(x) := \int_{-\infty}^{+\infty} f(x - \xi)g(\xi)d\xi \tag{4.34}$$

is well defined and

$$\widehat{f * g} = \hat{f}\hat{g}. \tag{4.35}$$

The proofs are left to the reader.

So far we have restricted ourselves to functions of fast decrease. There are less well behaved functions whose Fourier transforms are well defined and share some of the above properties. As these functions are special cases of the distributions discussed in the next section, we only indicate two examples here.

The first example is related to the diffraction by a slit which verifies the uncertainty principle experimentally. Consider the transmission function of an ideal slit with "width" $2a$

$$f(x) = \begin{cases} 1, & |x| < a \\ 0, & |x| > a. \end{cases} \tag{4.36}$$

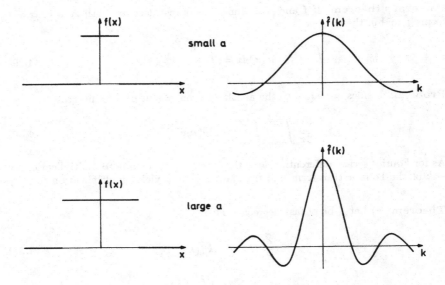

Fig. 4.4 Diffraction by a slit and uncertainty principle

According to Huygens' principle the interference pattern of the diffracted beam far away from the slit is described by the Fourier transform of f.

$$\hat{f}(k) = \frac{1}{\sqrt{2\pi}} \int_{-a}^{+a} e^{ikx} dx = \frac{1}{\sqrt{2\pi}ik} e^{ikx} \Big|_{-a}^{+a}$$

$$= \sqrt{\frac{2}{\pi}} \frac{1}{2i} \frac{e^{ika} - e^{-ika}}{k} = \sqrt{\frac{2}{\pi}} \frac{\sin ka}{k}. \tag{4.37}$$

Note that in this example the integral of Fourier's inversion formula is not absolutely convergent and equation (4.31) does not hold in the sense of functions.

In the second example we calculate the Fourier transform of the Breit-Wigner function

$$f(x) \doteq \frac{a}{x^2 + a^2}, \quad a > 0. \tag{4.38}$$

$$\hat{f}(k) = \frac{a}{\sqrt{2\pi}} \int_{-\infty}^{+\infty} \frac{e^{ikx}}{x^2 + a^2} dx = \frac{a}{\sqrt{2\pi}} \int_{-\infty}^{+\infty} \frac{e^{ikx}}{(x + ia)(x - ia)} dx. \tag{4.39}$$

For positive k we use the residue theorem with the closed curve of Fig. 4.5 to evaluate the integral.

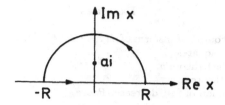

Fig. 4.5   The curve C in the complex x-plane

$$\int_C \frac{e^{ikz}}{z^2+a^2}dz = \int_{-R}^{+R}\frac{e^{ikz}}{x^2+a^2}dx + \int_{half\ circle}\frac{e^{ikz}}{z^2+a^2}dz$$

$$= 2\pi i \mathrm{Res}\frac{f}{g}\Big|_{ia} = 2\pi i\frac{f(ia)}{g'(ia)} = \frac{\pi}{a}e^{-ka}$$

(4.40)

according to the quotient theorem of section 1.6 with

$$f(ia) = e^{-ka} \neq 0, \tag{4.41}$$

$$g(ia) = 0, \tag{4.42}$$

$$g'(ia) = 2ia \neq 0. \tag{4.43}$$

Equation (4.40) holds for arbitrary positive R. In the limit as R tends to infinity the integral over the half circle tends to zero by the argument following equations (1.101) because e^{ikz} is bounded in the upper half plane if k is positive. Therefore

$$\hat{f}(k) = \sqrt{\frac{\pi}{2}}e^{-ka}, \quad k < 0. \tag{4.44}$$

For negative k the integration must be carried out over the lower half circle enclosing the isolated singularity at $-ia$ and yields

$$\hat{f}(k) = \sqrt{\frac{\pi}{2}}e^{ka}, \quad k < 0. \tag{4.45}$$

For vanishing k we have already calculated the Fourier transform in example b of section 1.7

$$\hat{f}(0) = \sqrt{\frac{\pi}{2}}, \tag{4.46}$$

altogether

$$\hat{f}(k) = \sqrt{\frac{\pi}{2}}e^{-|k|a}. \tag{4.47}$$

Note that $\hat{f}$ is not differentiable as a function and this time equation (4.32) fails.

Exercises

1. Calculate the Fourier transform of $f(x) = x e^{-x^2/a^2}$ by
   a) completing squares
   b) using the last theorem.
2. Prove the last theorem.
3. Let f be a function of fast decrease. Prove

$$\hat{\hat{f}}(x) = f(-x).$$

4. Let f be a real function of fast decrease. Prove

$$\hat{f}(-x) = \overline{\hat{f}}(k).$$

5. Fourier transform the transmission function of a double slit.
6. Compute the Fourier transform of

$$f(x) := \begin{cases} x, & -1 < x < 1 \\ 0, & \text{elsewhere.} \end{cases}$$

7. Does Parseval's theorem hold for the functions (4.36) and (4.38)?

4.3 The Fourier transform of distributions

By generalizing the Fourier transform to distributions we shall avoid the difficulties met in the last two examples. However, defining the Fourier transform of distributions is slightly more complicated than the Fourier series of periodic distributions because in the non-periodic case harmonics are not test functions any more. Consequently we can define the Fourier transform only for a certain class of distributions. These are called tempered distributions and defined by requiring all functions in the definition of the distribution to be of fast decrease.

Definition: A tempered distribution $D(x)$ is an equivalence class of sequences of functions $f_n(x)$ all of fast decrease such that for any function $S(x)$ also of fast decrease the sequence of numbers $\int_{-\infty}^{+\infty} f_n(x) S(x) dx$ converges. As before we denote its limit by

$$\lim_{n \to \infty} \int_{-\infty}^{+\infty} f_n(x) S(x) dx =: \int_{-\infty}^{+\infty} D(x) S(x) dx =: D(S). \qquad (4.48)$$

Examples: a) The δ-distribution is tempered. Indeed let

$$f_n(x) := \frac{n}{\sqrt{\pi}} e^{-n^2 x^2} \qquad (4.49)$$

and let $S(x)$ be any function of fast decrease. Then

$$\left| \int_{-\infty}^{+\infty} \frac{n}{\sqrt{\pi}} e^{-n^2 x^2} S(x) - S(0) \right|$$

$$= \left| \int_{-\infty}^{+\infty} \frac{n}{\sqrt{\pi}} e^{-n^2 x^2} [S(x) - S(0)] dx \right|$$

$$\leq \int_{-\infty}^{+\infty} \frac{n}{\sqrt{\pi}} e^{-n^2 x^2} \left| S(x) - S(0) \right| dx \tag{4.50}$$

$$\leq \max \left| S'(x) \right| \int_{-\infty}^{+\infty} \frac{n}{\sqrt{\pi}} e^{-n^2 x^2} |x| dx$$

$$= \max |S'(x)| \frac{1}{\sqrt{\pi} n} \xrightarrow[n \to \infty]{} 0$$

In general every distribution of compact support is tempered.

b) The constant function 1 is a tempered distribution. Let

$$f_n(x) := e^{-x^2/n^2}. \tag{4.51}$$

Then

$$\left| \int_{-\infty}^{+\infty} e^{-x^2/n^2} S(x) dx - \int_{-\infty}^{+\infty} 1 S(x) dx \right|$$

$$\leq \int_{-\infty}^{+\infty} \frac{1 - e^{-x^2/n^2}}{1 + x^2} \left| (1 + x^2) S(x) \right| dx \tag{4.52}$$

$$\leq \max \left| (1 + x^2) S(x) \right| \int_{-\infty}^{+\infty} \frac{1 - e^{-x^2/n^2}}{1 + x^2} dx \xrightarrow[n \to \infty]{} 0.$$

A differentiable function defined on the entire x-axis is said to be of slow increase if its increase at plus and minus infinity is polynomial or slower. E. g. a polynomial or $e^{-1/x^2} \ln x^2$ is a function of slow increase. Any tempered distribution multiplied by a function of slow increase is still tempered. In particular any function of slow increase is tempered. On the other hand the function e^x is not a tempered distribution. Let us note two more properties of tempered distributions. Any linear combination of tempered distributions is tempered. The derivative of a tempered distribution is tempered.

We are now ready to define the Fourier transform of a tempered distribution by Fourier transforming its representative functions.

Definition: Let $D(x)$ be a tempered distribution represented by the functions $f_n(x)$ of fast decrease. By Parseval's theorem their Fourier transforms $\hat{f}_n(k)$ define a tempered distribution.

$$\lim_{n \to \infty} \int_{-\infty}^{+\infty} \hat{f}_n(k) S(k) dk = \lim_{n \to \infty} \int_{-\infty}^{+\infty} f_n(x) \hat{S}(x) dx. \tag{4.53}$$

By definition this distribution is the Fourier transform of $D(x)$ and we denote it by $\hat{D}(k)$.

Examples: The Fourier transform of the δ-distribution concentrated at the origin is the constant function $1/\sqrt{2\pi}$. Indeed by equation (4.16) the Fourier transform of the sequence (4.49) representing the δ-distribution is

$$\hat{f}_n(k) = \frac{n}{\sqrt{\pi}} \frac{1}{\sqrt{2n}} e^{-k^2/(4n^2)} \underset{n\to\infty}{\longrightarrow} \frac{1}{\sqrt{2\pi}}. \tag{4.54}$$

In this example the uncertainty principle is pushed to its extreme limit, the width of the δ-distribution is zero, the width of its Fourier transform is infinite. Likewise the Fourier transform of the "plane wave"

$$f(x) = \frac{1}{\sqrt{2\pi}} e^{-ik_0 x} \tag{4.55}$$

regarded as tempered distribution is the δ-distribution concentrated at k_0

$$\hat{f}(k) = \delta(k - k_0). \tag{4.56}$$

This formula also allows us to consider now the Fourier series as a particular case of the Fourier transformation. In fact any distribution $D(x)$ of period L over the real line is tempered and its Fourier transform follows from its Fourier series:

$$D(x) = \sum_{j=-\infty}^{+\infty} c_j e^{-ij2\pi x/L}, \tag{4.57}$$

$$\hat{D}(k) = \frac{1}{\sqrt{2\pi}} \sum_{j=-\infty}^{+\infty} c_j \delta(k - j\frac{2\pi}{L}x). \tag{4.58}$$

This distribution is tempered because the Fourier ceofficients c_j are of slow increase in j.

Some of the theorems concerning Fourier transforms of functions of fast decrease generalize immediately to tempered distributions.

Theorem: For any tempered distribution $D(x)$ we have the Fourier inversion formula under the form

$$\hat{\hat{D}}(x) = D(-x). \tag{4.59}$$

Theorem: a) Let $D(x)$ be tempered. Then

$$\frac{\widehat{dD}}{dx}(k) = -ik\hat{D}(k), \tag{4.60}$$

$$(\widehat{ixD})(k) = \frac{d\hat{D}}{dk}(k). \tag{4.61}$$

b) If $D(x) = E(x - a)$, then

$$\hat{D}(k) = e^{ika}\hat{E}(k). \tag{4.62}$$

c) If D is even (odd), then so is its Fourier transform.

Example:

$$\widehat{\delta'(k)} = \frac{-ik}{\sqrt{2\pi}}. \tag{4.63}$$

Both distributions are odd.

As distributions cannot be multiplied in general, Parseval's theorem only holds in the weaker form already used to define $\hat{D}$:

$$\int_{-\infty}^{+\infty} \hat{D}(k)S(k)dk = \int D(x)\hat{S}(x)dx \tag{4.64}$$

where D is a tempered distribution, S a function of fast decrease. Note the analogy with equation (3.66) for a periodic distribution. However, contrary to the case of periodic distributions,

$$\frac{1}{\sqrt{2\pi}}\int_{-\infty}^{+\infty} D(x)e^{-ikx}dx \tag{4.65}$$

makes no sense for a general tempered distribution, nor can we write the inversion formula in integral form. There are some exceptions worth mentioning:

a) If $D(x)$ is a distribution of compact support, then the above expression makes sense and yields the Fourier transform $\hat{D}(k)$, for instance

$$\hat{\delta}_{x_0}(k) = \frac{1}{\sqrt{2\pi}}\int_{-\infty}^{+\infty} \delta_{x_0}(x)e^{-ikx}dx = \frac{1}{\sqrt{2\pi}}e^{-ikx_0}. \tag{4.66}$$

b) **Theorem:** Every (absolutely) integrable function $f(x)$, i.e. a function satisfying

$$\int_{-\infty}^{+\infty} \left|f(x)\right|dx < \infty, \tag{4.67}$$

is a tempered distribution. Its Fourier transform is again a function, not necessarily integrable, which may be calculated by

$$\hat{f}(k) = \frac{1}{\sqrt{2\pi}}\int_{-\infty}^{+\infty} f(x)e^{ikx}dx. \tag{4.68}$$

The two functions at the end of the last section are examples of this theorem. Note, however, that the inversion formula does not hold in its integral form (4.22) for an arbitrary integrable function $f(x)$.

c) **Theorem:** Every square integrable function $f(x)$, i.e. a function satisfying

$$\int_{-\infty}^{+\infty} |f(x)|^2 dx < \infty, \tag{4.69}$$

is a tempered distribution. Its Fourier transform $\hat{f}(k)$ is again a square integrable function.

Now, however, both equation (4.68) and the inversion formula (4.22) may fail in the sense of functions. For example $f(x) = (1 + x^2)^{-1/2}$ is square integrable but

$$\hat{f}(0) = \int_{-\infty}^{+\infty} \frac{dx}{\sqrt{1 + x^2}} \tag{4.70}$$

diverges logarithmically. By Abel's theorem $\hat{f}(k)$ is well defined for all other values of k, in this example. For square integrable functions the following theorem makes precise the link between Fourier series and Fourier transform hinted at in the introduction:

Theorem: Let $f(x)$ be square integrable. Then the sequence of functions

$$\hat{f}_L(k) := \frac{1}{\sqrt{2\pi}} \int_{-L/2}^{+L/2} f(x) e^{ikx} dx \tag{4.71}$$

converges in the mean to $\hat{f}(k)$ for L going to infinity and

$$f_M(x) := \frac{1}{\sqrt{2\pi}} \int_{-M/2}^{+M/2} \hat{f}(k) e^{-ikx} dk \tag{4.72}$$

converges in the mean to $f(x)$ for M going to infinity.

Exercises

1. Define distributions of compact support and show that they are tempered.
2. Show that $e^{-1/x^2} \ln x^2$ is of slow increase.
3. Verify equation (4.53).
4. Indicate a sequence of functions that represents $f(x) = x^2$ as tempered distribution.
5. Fourier transform

    a) x^4 b) $\delta''(x - 2)$
    c) $\sin^2 x$ d) the harmonic saw.

4.4 Calculating a Green function

We are now ready for the promised algorithmic derivation of the Green function of the harmonic oscillator

$$\left[\left(\frac{d}{dt}\right)^2 + 2\lambda\frac{d}{dt} + \omega_0^2\right] G(t) = \delta(t), \quad \omega_0 > \lambda > 0. \tag{4.73}$$

Assuming $G(t)$ tempered, we Fourier transform this differential equation to get an algebraic equation in frequency space which is easy to solve. We back transform the solution using the residue theorem and obtain the Green function. The first step yields

$$\left[(-i\omega)^2 - 2i\lambda\omega + \omega_0^2\right]\hat{G}(\omega) = \frac{1}{\sqrt{2\pi}} \tag{4.74}$$

with solution

$$\hat{G}(\omega) = \frac{1}{\sqrt{2\pi}}\frac{1}{-\omega^2 - 2i\lambda\omega - \omega_0^2}. \tag{4.75}$$

This function is square integrable so that $G(t)$ is indeed tempered. It is also integrable. Therefore the inversion formula can be written as an integral.

$$G(t) = \frac{1}{2\pi}\int_{-\infty}^{+\infty}\frac{e^{-i\omega t}d\omega}{-\omega^2 - 2i\lambda\omega + \omega_0^2} = \frac{-1}{2\pi}\int_{-\infty}^{+\infty}\frac{e^{-i\omega t}d\omega}{(\omega - \omega_1)(\omega - \omega_2)}. \tag{4.76}$$

The integrand has isolated singularities at

$$\omega_1 := -i\lambda - \sqrt{\omega_0^2 - \lambda^2}, \tag{4.77}$$

$$\omega_2 := -i\lambda + \sqrt{\omega_0^2 - \lambda^2}. \tag{4.78}$$

For negative t we evaluate the integral with Cauchy's theorem and the closed curve of fig. 4.6.

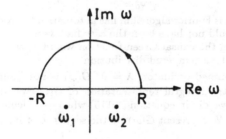

Fig. 4.6 Closed curve for negative times

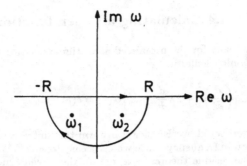

Fig. 4.7   Closed curve for positive times

In the limit R tending to infinity the integral over the half circle vanishes and we obtain

$$G(t) = 0, \quad t < 0. \tag{4.79}$$

For positive times we must use the clockwise oriented curve in the lower half plane, fig. 4.7 and accordingly the residue theorem applies:

$$G(t) = \frac{-2\pi i}{2\pi} \left[\text{Res}\Big|_{\omega_1} + \text{Res}\Big|_{\omega_2} \frac{e^{-i\omega t}}{-\omega^2 + 2i\lambda\omega - \omega_0^2} \right]. \tag{4.80}$$

Using the quotient theorem we find:

$$\begin{aligned}
G(t) &= -i \left[\frac{e^{-i\omega_1 t}}{-2\omega_1 - 2i\lambda} + \frac{e^{-i\omega_2 t}}{-2\omega_2 - 2i\lambda} \right] \\
&= -ie^{-\lambda t} \left[\frac{e^{i\sqrt{\omega_0^2 - \lambda^2}t}}{2\sqrt{\omega_0^2 - \lambda^2}} - \frac{e^{-i\sqrt{\omega_0^2 - \lambda^2}t}}{2\sqrt{\omega_0^2 - \lambda^2}} \right] \\
&= \frac{1}{\sqrt{\omega_0^2 - \lambda^2}} e^{-\lambda t} \sin\sqrt{\omega_0^2 - \lambda^2}t, \quad t > 0.
\end{aligned} \tag{4.81}$$

All together

$$G(t) = \frac{H(t)}{\sqrt{\omega_0^2 - \lambda^2}} e^{-\lambda t} \sin\sqrt{\omega_0^2 - \lambda^2}t. \tag{4.82}$$

Let us note that this Fourier algorithm has automatically produced the causal Green function. This would not have been the case had we allowed (unphysical) negative values for λ. Then the causal Green function would grow exponentially with time and would fail to be a tempered distribution.

For the undamped oscillator, $\lambda = 0$, $\hat{G}(\omega)$ is not integrable. In order to be able to still use the integral representation of $G(t)$, we must "regularize" $\hat{G}(\omega)$ similarly to what we did in equation (2.115) when we calculated the Green function of electrostatics. We represent $\hat{G}(\omega)$ by the sequence of integrable functions

$$\hat{G}_n(\omega) := \frac{1}{\sqrt{2\pi}} \frac{1}{-\omega^2 - 2i\omega/n + \omega_0^2}. \tag{4.83}$$

The Green function is then given by the limit of equation (4.82) as λ tends to zero.

$$G(t) = \frac{H(t)}{\omega_0} \sin \omega_0 t. \qquad (4.84)$$

The Fourier algorithm outlined here for the damped harmonic oscillator works for many other linear (ordinary and partial) differential equations with constant coefficients. Some prominent examples from physics will be discussed in chapter 5.

Exercises

1. Derive the Green function of the over-damped harmonic oscillator $\lambda > \omega_0 > 0$.
2. Verify that equation (4.84) solves the differential equation (4.73) with $\lambda = 0$.
3. Consider the Schrödinger equation of the harmonic oscillator

$$\left[-\left(\frac{d}{dx}\right)^2 + x^2 \right] \psi(x) = E\psi(x).$$

What differential equation is satisfied by $\hat{\psi}(k)$?

4.5 Fourier transforms in higher dimensions

The generalization of Fourier transforms to functions and distributions over $\mathbb{R}^N$ is straight forward. All definitions and theorems of sections 4.2 and 4.3 remain valid upon replacement of single integrals by multiple integrals, replacement of the factors $(2\pi)^{-\frac{1}{2}}$ in front of single integrals by $(2\pi)^{-\frac{N}{2}}$ in front of multiple integrals and replacement of ordinary derivatives by partial derivatives. E. g. let $f : \mathbb{R}^N \to \mathbb{C}$ be a function of fast decrease. Its Fourier transform is

$$\hat{f}(k^1, ..., k^N) := \frac{1}{\sqrt{2\pi}^N} \int_{-\infty}^{+\infty} ... \int_{-\infty}^{+\infty} f(x^1, ..., x^N) e^{ik^1 x^1} ... e^{ik^N x^N} dx^1 ... dx^N$$

$$= \frac{1}{\sqrt{2\pi}^N} \int f(x) e^{ik \cdot x} dx$$

$$(4.85)$$

where $k \cdot x$ denotes the Euclidean scalar product on $\mathbb{R}^N$:

$$k \cdot x := \sum_{j=1}^{N} k^j x^j. \qquad (4.86)$$

Similarly the back transform becomes

$$f(x) = \frac{1}{\sqrt{2\pi}^N} \int \hat{f}(k) e^{-ik \cdot x} dk. \qquad (4.87)$$

Instead of the Euclidean metric we may as well choose a pseudometric, for instance if we denote by $x = (x^0, x^1, ...x^N)$ and $k = (k^0, k^1, ...k^N)$ elements in Minkowski space

$$k \cdot x := k^0 x^0 - \sum_{j=1}^{N} k^j x^j =: \sum_{\mu,\nu=0}^{N} k^\mu \eta_{\mu\nu} x^\nu. \qquad (4.88)$$

The pseudometric simply produces the "wrong" sign convention of the Fourier transform in the "space-like" coordinates $x^1, x^2, ..., x^N$.

In the higher dimensional case k is also referred to as wave vector. Independently of the choice of metric, k may be interpreted as element of the dual space of the vector space $\mathbb{R}^N$ in which x lives. As another side remark we note that already functions of fast decrease and tempered distributions are only defined on the entire (vector space) $\mathbb{R}^N$ unlike test functions and general distributions which can be defined on subsets Ω and also on manifolds.

In higher dimensions partial differentiation in x-space becomes multiplication by the corresponding component in k-space:

$$\widehat{\frac{\partial}{\partial x^j} f}(x) = -\mathrm{i} k^j \hat{f}(k), \qquad j = 1, 2, ..., N \qquad (4.89)$$

and vice versa

$$\frac{\partial}{\partial k^j} \hat{f}(k) = \mathrm{i} \widehat{x^j f}(x), \qquad j = 1, 2, ..., N. \qquad (4.90)$$

With the choice of a Lorentz metric (4.88), however, there are additional minus signs:

$$\widehat{\frac{\partial}{\partial x^\mu} f}(k) = -\mathrm{i} \eta_{\mu\mu} k^\mu \hat{f}(k) = -\mathrm{i} \sum_{\nu=0}^{N} \eta_{\mu\nu} k^\nu \hat{f}(k), \quad \mu = 0, 1, ..., N, \qquad (4.91)$$

$$\frac{\partial}{\partial k^\mu} \hat{f}(k) = \mathrm{i} \eta_{\mu\mu} \widehat{x^\mu f}(x) = \mathrm{i} \sum_{\nu=0}^{N} \eta_{\mu\nu} \widehat{x^\nu f}(x). \qquad (4.92)$$

There is one new property in higher dimensions: The Fourier transform of a tensor product of two tempered distributions $D(x)$ and $E(y)$ equals the tensor product of the Fourier transforms.

$$\widehat{(D \otimes E)}(k, l) = \hat{D}(k) \otimes \hat{E}(l) \qquad (4.93)$$

with $x, k \in \mathbb{R}^N$ and $y, l \in \mathbb{R}^M$. For example

$$f(x^1, .., x^n) := \exp[\frac{-1}{a^2} \sum_{j=1}^{N} (x^j)^2], \quad a > 0, \qquad (4.94)$$

$$\hat{f}(k^1, .., k^n) = (\frac{a}{\sqrt{2}})^N \exp[-\frac{a^2}{4} \sum_{j=1}^{N} (k^j)^2] \qquad (4.95)$$

and therefore

$$\hat{\delta}(k) = 1/\sqrt{2\pi}^N, \quad k \in \mathbb{R}^N. \qquad (4.96)$$

We close this section with a result useful for calculating Green functions of differential operators with spherical symmetry.

Theorem: The function on $\mathbb{R}^N$

$$f(x) = r^a \qquad (4.97)$$

with $r := \left[\sum_{j=1}^{N} (x^j)^2\right]^{\frac{1}{2}}$ and with $-N < a < 0$ is a tempered distribution. Its Fourier transform is

$$\hat{f}(k) = 2^{a+N/2} \frac{\Gamma(a/2 + N/2)}{\Gamma(-a/2)} q^{-a-N} \qquad (4.98)$$

with $q := \left[\sum_{j=1}^{N} (k^j)^2\right]^{\frac{1}{2}}$. Let us recall a few values of the gamma function $\Gamma(a)$:

$$\Gamma(1) = 1, \qquad (4.99)$$

$$\Gamma(\frac{1}{2}) = \sqrt{\pi}, \qquad (4.100)$$

$$\Gamma(a + 1) = a\Gamma(a). \qquad (4.101)$$

The proof of the theorem uses the regularization of equation (2.115) and N-dimensional polar coordinates. We skip the details.

Exercises

1. Fourier transform the transmission function of a rectangular window

$$f(x^1, x^2) = \begin{cases} 1 & \text{if } |x^1| < a \quad \text{and} \quad |x^2| < b \\ 0 & \text{elsewhere} \end{cases}$$

2. Verify equations (4.91) and (4.92).
3. Compute the Fourier transforms (4.95) and (4.96) with respect to the Euclidean and Lorentz metric.

4.6 The Laplace transform

We end this chapter with a few remarks about a further generalization of Fourier transforms. The Fourier transformation (in $N = 1$ dimension) associates to a function $f(x)$ (or distribution) of one real variable and with complex values a function $\hat{f}(k)$ of the same type. In some practical calculations we have analytically continued $\hat{f}(k)$ to complex arguments

$$z := \lambda - ik. \tag{4.102}$$

Note that here due to convention minus k is the imaginary part of z. The resulting complex function $\hat{f}(z)$ is called the Laplace transform $(\mathcal{L}f)(z)$ of $f(x)$:

$$
\begin{aligned}
(\mathcal{L}f)(z) &:= \frac{1}{\sqrt{2\pi}} \int_{-\infty}^{+\infty} f(x)e^{-zx}dx \\
&= \frac{1}{\sqrt{2\pi}} \int_{-\infty}^{+\infty} e^{-\lambda x} f(x) e^{ikx} dx = e^{-\lambda x} \widehat{f}(x)(k).
\end{aligned}
\tag{4.103}
$$

Due to the exponential factor $e^{-\lambda x}$ the Fourier transform of $e^{-\lambda x} f(x)$ sometimes fails to exist for $\lambda \neq 0$ even if $f(x)$ is of fast decrease. For every distribution $D(x)$ let us denote by I_D the interval consisting of values λ such that $e^{-\lambda x} D(x)$ is tempered. For example if $D(x) = e^{-x^2}$, this interval is the entire λ-axis, $I_{e^{-x^2}} = \mathbb{R}$, if $D(x) = e^{x^2}$ the interval is empty $I_{e^{x^2}} = \emptyset$ and for the step function $D(x) = H(x)$ the interval is the half axis $I_{H(x)} = [0, \infty)$.

An inconvenience of the Laplace tranformation is that we have to deal with functions of variable domain of definition $z = \lambda - ik \in I_D \times \mathbb{R}$. Its advantages are that the Laplace transformation can also be applied to certain non-tempered distributions and more importantly that it leads to holomorphic functions.

Theorem: Let $D(x)$ be a distribution. If the interior $\overset{\bullet}{I}_D$ of the interval I_D is non-empty, then the Laplace transform of D

$$(\mathcal{L}D)(z) := e^{-\lambda x} \widehat{D}(x)(k) \tag{4.104}$$

is a holomorphic function over the strip in the complex plane defined by $z = \lambda - ik \in \overset{\bullet}{I}_D \times \mathbb{R}$.

For example if $D(x) = \delta_{x_0}(x)$, $\overset{\bullet}{I}_D = \mathbb{R}$ and its Laplace transform

$$
\begin{aligned}
(\mathcal{L}\delta_{x_0})(z) &:= e^{-\lambda x}\widehat{\delta_{x_0}}(k) = e^{-\lambda x_0} \hat{\delta}_{x_0}(k) \\
&= \frac{1}{\sqrt{2\pi}} e^{-\lambda x_0} e^{ikx_0} = \frac{1}{\sqrt{2\pi}} e^{-x_0 z}.
\end{aligned}
\tag{4.105}
$$

is holomorphic everywhere.

If $D(x) = x^j H(x)$ with $j = 0, 1, 2...$, then $\overset{\circ}{I}_D = (0, \infty)$ and the Laplace transform

$$(\mathcal{L}x^j H(x))(z) = \frac{j!}{\sqrt{2\pi}} z^{-j-1} \tag{4.106}$$

is holomorphic in the complex half plane of positive real part λ. Note that we have already encountered other particular cases of this theorem , e. g. equation (4.75).

A function $f(x)$ may be retrieved from its Laplace transform $(\mathcal{L}f)(z)$ using the following convenient form of the inversion formula which generalizes the trick of completing squares:

If for some $\lambda \in I_f$ the function $e^{-\lambda x} f(x)$ is of fast decrease, then

$$f(x) = \frac{-i}{\sqrt{2\pi}} \int_{\lambda-i\infty}^{\lambda+i\infty} (\mathcal{L}f)(z) e^{xz} dz \tag{4.107}$$

where the complex integration is over the vertical line with real part λ.

In conclusion, the Laplace transformation formalizes the integration tricks of chapter one in the context of Fourier integrals. Furthermore, as in the case of the undamped oscillator, the Laplace transformation automatically produces necessary regularization factors, i. e. the damping term $e^{-\lambda t}$.

Exercise

1. Calculate the Laplace transform (4.106).

5

Some linear differential operators

and their Green functions

This chapter is concerned with a few linear partial differential operators with constant coefficients. We restrict ourselves to operators of second order:

$$A := \sum_{j,\ell=1}^{N} a_{j\ell} \frac{\partial^2}{\partial x^j \partial x^\ell} + \sum_{j=1}^{N} b_j \frac{\partial}{\partial x^j} + c. \tag{5.1}$$

The coefficients $a_{j\ell}$, b_j and c are complex numbers. If $T(x)$ is a tempered distribution, then the Fourier transform of AT is given by

$$\widehat{AT}(k) = P(k)\hat{T}(k) \tag{5.2}$$

where

$$P(k) := -\sum_{j,\ell=1}^{N} a_{j\ell} k^j k^\ell - i \sum_{j=1}^{N} b_j k^j + c \tag{5.3}$$

is called the polynomial of A. Note that in equation (5.2) we understand the Fourier transform with respect to the Euclidean scalar product $k \cdot x$ in $\mathbb{R}^N$. Before passing to specific operators of interest to physicists, we mention a general result.

Theorem: Let A be a linear differential operator with constant coefficients and $s(x)$ a tempered distribution. Then the inhomogeneous differential equation

$$AT = s \tag{5.4}$$

is solved by at least one tempered distribution $T(x)$. If the polynomial of A has no real roots, the tempered solution T is unique.

Since the δ-distribution is tempered the theorem implies that every linear differential operator with constant coefficients has a tempered Green function. By the theorem on Laplace transforms this Green function can never be of compact support.

5.1 The Laplace equation

The Laplace operator

$$\Delta := \left(\frac{\partial}{\partial x^1}\right)^2 + \dots + \left(\frac{\partial}{\partial x^N}\right)^2 \tag{5.5}$$

appears in many branches of physics, e.g. in elasticity theory, hydrodynamics, electrostatics. The solutions of the homogeneous Laplace equation $\Delta D = 0$ are functions. They are called harmonic. More generally, if $s(x)$ is a differentiable (or analytic) function, then every distribution D solving $\Delta D = s$ is itself a differentiable (or analytic) function. We have already encountered harmonic functions of two variables in chapter 1. Every real function $f(x, y)$ is harmonic if and only if it is the real part of a holomorphic function of the complex variable $z = x + iy$. The corollary (1.91) states that the values of a holomorphic function inside a given closed curve are already determined by its values on this curve. This statement generalizes to harmonic functions in $\mathbb{R}^N$.

Theorem: Let Ω be an open subset of $\mathbb{R}^N$ with finite volume and $f : \bar{\Omega} \to \mathbb{C}$ a continuous function, where $\bar{\Omega}$ denotes the closure of Ω. If f is harmonic on Ω, then it is uniquely determined by its values on the boundary $\Sigma := \bar{\Omega} - \Omega$.

For $N = 1$ every harmonic function is linear and the theorem is obvious.

Now let us turn to the Green functions of the Laplace operator. We have to solve the inhomogeneous equation

$$\Delta G(x) = \delta(x). \tag{5.6}$$

Of course the solution is not unique, the solutions differ by harmonic functions. To obtain a particular solution for $N \geq 3$, we Fourier-transform equation (5.6)

$$-q^2 \hat{G}(k) = (2\pi)^{-N/2}. \tag{5.7}$$

Let us recall that q is the radius in Fourier space. Then we solve for $\hat{G}$

$$\hat{G}(k) = -(2\pi)^{-N/2} q^{-2} \tag{5.8}$$

and use the theorem of section 4.5

$$\begin{aligned}
G(x) = \hat{\hat{G}}(x) &= -(2\pi)^{-N/2} 2^{-2+N/2} \frac{\Gamma(-1+N/2)}{\Gamma(1)} r^{2-N} \\
&= \frac{-1}{N-2} \frac{1}{2} \pi^{-N/2} \Gamma(N/2) r^{2-N}.
\end{aligned} \tag{5.9}$$

Therefore a Green function of the Laplace operator in dimension $N \geq 3$ is given by

$$G(x) = \frac{-1}{N-2} \frac{1}{S_N} \frac{1}{r^{N-2}} \tag{5.10}$$

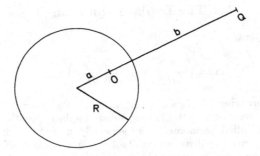

Fig. 5.1   Placing an image charge

where

$$S_N := \frac{2\pi^{N/2}}{\Gamma(N/2)} \tag{5.11}$$

is the area of the unit sphere in N dimensions. This Green function is characterized by its fall off at infinity. In three dimensions for instance

$$G(x) = \frac{-1}{4\pi r} \tag{5.12}$$

is the electric potential of a static point-like "unit" charge at the origin in an otherwise empty space.

If we are interested in the potential of the charge placed inside a closed conducting surface Σ, then we have to adjust the boundary condition on Σ. We must add to the Green function (5.12) the real function $f(x)$ determined by

$$\Delta f(x) = 0 \quad \text{for all} \quad x \in \Omega, \tag{5.13}$$

$$G(x) + f(x) = 0 \quad \text{for all} \quad x \in \Sigma. \tag{5.14}$$

It is difficult to find an explicit expression for this function unless the surface is highly symmetric. For the sphere for instance there is a well-known trick: placing an image charge. The image charge is a second point-like charge of value

$$Q = -\frac{R}{a} \tag{5.15}$$

placed outside of the sphere on the ray from the centre of the sphere through the unit charge at the origin. We denote by R the radius of the sphere and by a the distance between its centre and the origin as indicated in fig. 5.1. If the image charge is located at a distance

$$b = \frac{R^2}{a} \tag{5.16}$$

from the sphere, then its potential is the harmonic function $f(x)$ over the ball $\bar{\Omega}$ such that $G + f$ vanishes on the sphere.

In general we are interested in the Green functions to solve the inhomogeneous equation

$$\Delta f = s \tag{5.17}$$

where s is a given source function. Let us suppose that we know all Green functions relative to a closed surface Σ

$$\Delta G_\xi(x) = \delta_\xi(x) \quad \text{for all} \quad \xi \in \Omega, \tag{5.18}$$

$$G_\xi(x) = 0 \quad \text{for all} \quad \xi \in \Omega, \ x \in \Sigma. \tag{5.19}$$

Note that these Green functions are not invariant under translations. If the support of the source is contained in Ω then the magic formula

$$f(x) := \int_\Omega G_\xi(x)s(\xi)d\xi \tag{5.20}$$

solves the inhomogeneous equation subject to the boundary condition

$$f(x) = 0 \quad \text{for all} \quad x \in \Sigma. \tag{5.21}$$

We mention a second application of the Green functions (5.18-19) i.e. to solve the boundary value problem or Dirichlet problem. Let Σ be sufficiently regular and g a continuous function on Σ. Then we look for the harmonic function f on Ω that coincides with g on Σ. It is given by the formula valid in any dimension $N \geq 2$

$$f(x) = \int_\Sigma \frac{\partial G_\xi(x)}{\partial N_\xi} g(\xi)d\sigma(\xi) \tag{5.22}$$

where $\partial/\partial N_\xi$ denotes the component of the gradient with respect to ξ that is perpendicular to Σ and directed outward, $d\sigma$ is a surface element in Σ. This formula generalizes equation (1.91) from two to N dimensions. Note, however, that the construction of the Green functions $G_\xi(x)$ already involves a Dirichlet problem, equations (5.18-19), so that the formula (5.22) is really useful only for surfaces Σ where the Green function $G_\xi(x)$ can be constructed by other means, e.g. the image charge for the sphere. Note also the similarity between the solution (5.22) of the boundary value problem and the solution (2.102) of the initial value problem of the harmonic oscillator. While the latter propagates in time, the former propagates boundary values into the interior Ω along directions perpendicular to the boundary.

We still need a two-dimensional Green function. For $N = 2$ the back transform (5.9) has to be evaluated separately yielding

$$G(x) = \frac{1}{2\pi} \log r. \tag{5.23}$$

For completeness we add an obvious Green function in one dimension: $G(x) = xH(x)$.

We close this section with an application in general relativity. As already mentioned, Einstein's equations are non-linear and do not allow for point-like particles. There is a remarkable exception, the gravitational field of a massless particle discovered in 1969 by Bonnor (1969). Imagine a massless particle in an otherwise empty space flying along the positive z-axis with energy density $\rho(x, y, z - ct)$, c the speed of light. Introducing light-cone coordinates

$$u := \frac{ct - z}{\sqrt{2}}, \tag{5.24}$$

$$v := \frac{ct - z}{\sqrt{2}}, \tag{5.25}$$

Bonnor makes the following ansatz for the metric:

$$d\tau^2 = 2dudv + 2A(x, y, u)du^2 - dx^2 - dy^2. \tag{5.26}$$

This ansatz reduces Einstein's equations to the two dimensional Laplace equation

$$\left(\frac{d^2}{dx^2} + \frac{d^2}{dy^2} \right) A = 16\pi G\rho \tag{5.27}$$

where G is Newton's constant. Therefore the massless particle with energy E admits the point-like limit

$$\rho(x, y, z - ct) = E\delta(x)\delta(y)\delta(z - ct) \tag{5.28}$$

with solution in the sense of distributions:

$$A(x, y, u) = \frac{4}{\sqrt{2}}GE \, \log(x^2 + y^2)\delta(u). \tag{5.29}$$

Exercises

1. Use the Laplace transform to show that no differential operator with constant coefficients can have a Green function of compact support.
2. Show that the Laplace operators in $\mathbb{R}^2$ and $\mathbb{R}^3$ take the following form in polar coordinates (2.116-118):

$$\Delta = \frac{\partial^2}{\partial r^2} + \frac{1}{r}\frac{\partial}{\partial r} + \frac{1}{r^2}\frac{\partial^2}{\partial \phi^2},$$

$$\Delta = \frac{\partial^2}{\partial r^2} + \frac{2}{r}\frac{\partial}{\partial r} + \frac{1}{r^2 \sin^2\theta}\frac{\partial^2}{\partial \phi^2} + + \frac{1}{r^2}\frac{\partial^2}{\partial \theta^2} + \frac{\cot\theta}{r^2}\frac{\partial}{\partial \theta}$$

and verify that the Green functions (5.23) and (5.12) are harmonic outside the origin.

3. Use a regularization similar to equation (2.103) to prove

$$\Delta \frac{\log r}{2\pi} = \delta(x^1, x^2).$$

4. Verify the trick of the image charge.
5. Show that in two dimensions the Green functions relative to the unit circle Σ around the origin are given by

$$G_\xi(x) = \frac{1}{4\pi} \log \frac{x \cdot x + y \cdot y - 2x \cdot y}{(x \cdot x)(y \cdot y) - 2x \cdot y + 1}.$$

Calculate $\partial G_\xi(x)/\partial N_\xi$ and show that equations (1.91) and (5.22) agree.

5.2 The heat equation

We now add dynamics to the Laplace operator and consider the following linear differential operator in $N + 1$ dimensions:

$$\frac{1}{a} \frac{\partial}{\partial t} - \Delta = \frac{1}{a} \frac{\partial}{\partial t} - \sum_{j=1}^{N} \left(\frac{\partial}{\partial x^j} \right)^2. \tag{5.30}$$

The corresponding equation

$$\left(\frac{1}{a} \frac{\partial}{\partial t} - \Delta \right) D(t, x) = s(t, x) \tag{5.31}$$

is called heat equation or diffusion equation. In the first case $D(t, x)$ is the temperature distribution in time and space, and the positive constant a is determined by specific heat and thermal conductivity of the medium. In the second case $D(t, x)$ is a particle density and a the diffusion coefficient. The rhs describes local production minus absorption. If a is purely imaginary,

$$a = \frac{\hbar}{2m} i \tag{5.32}$$

the homogeneous equation (5.31) is the free Schrödinger equation of a particle with mass m, $\hbar$ is Planck's constant. In the presence of a potential, the Schrödinger equation is still linear but with variable coefficients, and the Green function is known only in very simple cases.

Let us prove that

$$G(t, x) = \frac{aH(t)}{\sqrt{4\pi at}^N} e^{-r^2/(4at)} \tag{5.33}$$

is a Green function of the heat equation. Since $G(t,x)$ is a tempered distribution in the variables x, we can take its Fourier transform with respect to x using equation (4.95)

$$\hat{G}(t,k) = \frac{aH(t)}{\sqrt{2\pi}^N} e^{-q^2 at}. \tag{5.34}$$

We differentiate with respect to time:

$$\begin{aligned}
\frac{\partial \hat{G}}{\partial t} &= \frac{aH(t)}{\sqrt{2\pi}^N} e^{-q^2 at}(-q^2 a) + \frac{a}{\sqrt{2\pi}^N} \delta(t) e^{-q^2 at} \\
&= -aq^2 \hat{G} + \frac{a}{\sqrt{2\pi}^N} \delta(t)
\end{aligned} \tag{5.35}$$

and backtransform into x-space by means of equations (4.89) and (4.96)

$$\frac{\partial}{\partial t} G(t,x) = a\Delta G(t,x) + a\delta(t,x). \tag{5.36}$$

This causal Green function is characterized by three properties:
- It vanishes for negative times.
- It solves the homogeneous heat equation for positive times.
- Its limit in the sense of distributions in the variables x as t tends to zero is

$$\lim_{t\to 0} G(t,x) = a\delta(x) \tag{5.37}$$

by equation (4.95).

Therefore we have the following physical picture of this Green function. In $N = 1$ dimension, for example, it describes the temperature evolution of a rod initially at zero temperature after it is heated for a short period at the origin. Or in $N = 3$ dimensions it describes an ink cloud obtained by placing a small plastic bag filled with ink in a swimming pool and puncturing it at $t = 0$.

As for the harmonic oscillator the Green function can be used to solve the initial value problem or Cauchy problem. Given a function g of x, the initial value at $t = 0$, we look for a function $f(t,x)$ at later times that solves the homogeneous heat equation and coincides with g at $t = 0$:

$$\left(\frac{1}{a}\frac{\partial}{\partial t} - \Delta\right) f(t,x) = 0, \quad t > 0, \tag{5.38}$$

$$f(0,x) = g(x). \tag{5.39}$$

Note that we do not need initial velocities, because the heat equation is only of first order in time. Again we suppose that we already know the solution f and define the distribution $D(t,x)$ equal to $f(t,x)$ for positive times, equal to zero for negative times:

$$D(t,x) = H(t)f(t,x). \tag{5.40}$$

Then

$$\left(\frac{1}{a}\frac{\partial}{\partial t} - \Delta\right) D(t,x) = \frac{1}{a}H(t)\frac{\partial f}{\partial t} + \frac{1}{a}\delta(t)f - H(t)\Delta f = \frac{1}{a}\delta(t)g(x). \tag{5.41}$$

The magic formula solves this equation yielding D as a convolution product of the Green function and the distributional source $1/a\delta(t)g(x)$:

$$
\begin{aligned}
D(t,x) &= G(t,x) * \frac{1}{a}\delta(t)g(x) \\
&= \int_{\mathbb{R}^{N+1}} G(t-\tau, x-\xi)\frac{1}{a}\delta(\tau)g(\xi)d\tau d\xi \\
&= \int_{\mathbb{R}^{N+1}} \frac{aH(t-\tau)}{\sqrt{4\pi a(t-\tau)}^N} e^{-|x-\xi|^2/(4a(t-\tau))}\frac{1}{a}\delta(\tau)g(\xi)d\tau d\xi \\
&= \frac{H(t)}{\sqrt{4\pi at}^N} \int_{\mathbb{R}^N} e^{-|x-\xi|^2/(4at)}g(\xi)d\xi.
\end{aligned} \tag{5.42}
$$

Again this convolution product does not converge for arbitrary functions g, it does, for example, if g is continuous and bounded and a positive. Then the function on the half space $t > 0$

$$f(t,x) := \frac{1}{\sqrt{4\pi at}^N} \int_{\mathbb{R}^N} e^{-|x-\xi|^2/(4at)}g(\xi)d\xi \tag{5.43}$$

is differentiable and solves the Cauchy problem as may be verified directly.

A major inconvenience with the heat equation and with the Schrödinger equation is its incompatibility with special relativity. Indeed they propagate signals with unbounded velocities. According to equation (5.33) ink particles are found everywhere in the pool immediately after the puncture.

We close this section with a piece of notation which has become standard in quantum field theory, but demands due respect. It is related to, but distinct from, the complete Fourier transform $\hat{G}(\omega, k)$ of the causal Green function (5.33) for the Schrödinger equation: $a = \hbar i/(2m)$. The main ingredient for causality is the step function $H(t)$. Using an integration trick from section 1.7, the step function can be written under the following form:

$$H(t) = \lim_{\epsilon \to 0+} \frac{-1}{2\pi i} \int_{-\infty}^{+\infty} \frac{e^{-i\omega t}d\omega}{\omega + i\epsilon}. \tag{5.44}$$

Then equation (5.34) becomes

$$
\begin{aligned}
\hat{G}(t,k) &= \frac{aH(t)}{\sqrt{2\pi}^N} e^{-q^2 at} = \\
&= \lim_{\epsilon \to 0+} \frac{-\hbar}{2m}(2\pi)^{-N/2-1} \int_{-\infty}^{+\infty} \frac{e^{-i(\omega + \hbar q^2/(2m))t}}{\omega + i\epsilon}d\omega \\
&= \lim_{\epsilon \to 0+} \frac{1}{\sqrt{2\pi}} \int_{-\infty}^{+\infty} \frac{\hbar/(2m)(2\pi)^{-N/2+1/2}}{\omega - \hbar/(2m)q^2 + i\epsilon}e^{-i\omega t}d\omega.
\end{aligned} \tag{5.45}
$$

This formula is then abbreviated as

$$\hat{G}(\omega, k) = \frac{1}{\sqrt{2\pi}^{N+1}} \frac{\hbar/(2m)}{\omega - \hbar/(2m)q^2 + i\epsilon} \tag{5.46}$$

where it is understood that the limit of positive ϵ tending to zero is only performed after the integration over ω. It is tempting to take the limit $\epsilon \to 0^+$ immediately in equation (5.46) before integrating. In fact this limit would obey

$$P(\omega, k)\hat{G}(\omega, k) = \sqrt{2\pi}^{N+1}. \tag{5.47}$$

However, its Fourier transform would not reproduce the causal Green function. It is the positive ϵ that ensures causality.

Exercises

1. For what complex numbers a is (5.33) a valid Green function?
2. Show directly that the function (5.43) solves the Cauchy problem for continuous and bounded initial value.
3. Prove the representation (5.44) of the step function.
4. What Green function do you obtain from equation (5.46) for negative ϵ?

5.3 The wave equation

We define the d'Alembert or wave operator in $N + 1$ dimensions by

$$\begin{aligned}
\Box &:= \frac{1}{c^2} \frac{\partial^2}{\partial t^2} - \Delta = \left(\frac{\partial}{\partial x^0}\right)^2 - \left(\frac{\partial}{\partial x^1}\right)^2 - \cdots - \left(\frac{\partial}{\partial x^N}\right)^2 \\
&= \sum_{\mu,\nu=0}^{N} \eta^{\mu\nu} \frac{\partial}{\partial x^\mu} \frac{\partial}{\partial x^\nu}
\end{aligned} \tag{5.48}$$

with

$$x^0 := ct. \tag{5.49}$$

Its homogeneous equation describes for example the dynamics of a vibrating string (N=1), of a drum skin or the surface of a lake (N=2), of a sound wave in air or an electromagnetic wave in vacuum ($N = 3$). Then c is the speed of sound or light. More generally, Maxwell's equations written in the Lorentz gauge

$$\sum_{\mu=0}^{3} \frac{\partial}{\partial x^\mu} A^\mu = 0 \tag{5.50}$$

take the form

$$\Box A^\mu = \frac{1}{c^2 \epsilon_0} j^\mu \tag{5.51}$$

where A^0 is the electric potential divided by c

$$A^0 = \frac{1}{c} V \tag{5.52}$$

and j^0 is the charge density times c

$$j^0 = c\rho. \tag{5.53}$$

Lorentz invariance of the d'Alembert operator is evident from the last equation (5.48). In addition we shall see that the wave equation propagates signals with velocities less than or equal to c. From now on we shall always use x^0 as time variable but denote it by t for convenience. In other words, we put $c = 1$.

Since the wave equation is of second order in time, we must specify initial "positions" and "velocities" in order to have a unique solution of the Cauchy problem:

$$\Box f(t, x) = s(t, x), \quad t > 0, \tag{5.54}$$

$$f(0, x) = g(x), \tag{5.55}$$

$$\frac{\partial}{\partial t} f(0, x) = h(x). \tag{5.56}$$

As for the harmonic oscillator this problem is solved using the causal Green function

$$
\begin{aligned}
D(t, x) &= H(t)f(t, x) = G(t, x) * [H(t)s(t, x) + \delta(t)h(x) + \dot{\delta}(t)g(x)] \\
&= \int_{\mathbb{R}^{N+1}} G(t - \tau, x - \xi)H(\tau)s(\tau, \xi)d\tau d\xi \\
&\quad + \int_{\mathbb{R}^N} G(t, x - \xi)h(\xi)d\xi + \int_{\mathbb{R}^N} \frac{\partial}{\partial t} G(t, x - \xi)g(\xi)d\xi.
\end{aligned} \tag{5.57}
$$

In contrast to the heat equation there is no simple formula for the Green function of the wave operator valid in arbitrary dimensions N and we shall discuss the one, two, and three dimensional case separately.

$\underline{N = 1}$

Here the Green function is found easily after introducing light-cone coordinates:

$$u := \frac{t - x}{\sqrt{2}}, \qquad v := \frac{t + x}{\sqrt{2}}, \tag{5.58}$$

$$t = \frac{v + u}{\sqrt{2}}, \qquad x = \frac{v - u}{\sqrt{v}}. \tag{5.59}$$

The chain rule of partial derivatives yields

$$\frac{\partial}{\partial u} = \frac{\partial t}{\partial u}\frac{\partial}{\partial t} + \frac{\partial x}{\partial u}\frac{\partial}{\partial x} = \frac{1}{\sqrt{2}}\left(\frac{\partial}{\partial t} - \frac{\partial}{\partial x}\right), \tag{5.60}$$

$$\frac{\partial}{\partial v} = \frac{\partial t}{\partial v}\frac{\partial}{\partial t} + \frac{\partial x}{\partial v}\frac{\partial}{\partial x} = \frac{1}{\sqrt{2}}\left(\frac{\partial}{\partial t} + \frac{\partial}{\partial x}\right) \tag{5.61}$$

and the wave operator takes the form

$$\left(\frac{\partial^2}{\partial t^2} - \frac{\partial}{\partial x^2}\right)f(t,x) = \left(\frac{\partial}{\partial t} - \frac{\partial}{\partial x}\right)\left(\frac{\partial}{\partial t} + \frac{\partial}{\partial x}\right)f(t,x)$$

$$= 2\frac{\partial}{\partial u}\frac{\partial}{\partial v}\tilde{f}(u,v) \tag{5.62}$$

with

$$\tilde{f}(u,v) = f(t(u,v), x(u,v)). \tag{5.63}$$

Therefore the most general solution of the homogeneous wave equation is

$$\tilde{f}(u,v) = f_R(u) + f_L(v) \tag{5.64}$$

or in the old coordinates

$$f(x,t) = f_R\left(\frac{t-x}{\sqrt{2}}\right) + f_L\left(\frac{t+x}{\sqrt{2}}\right), \tag{5.65}$$

a superposition of two waves with unchanged profile (solitons), one moving right, the other moving left with velocity c. In light-cone coordinates the causal Green function is guessed easily:

$$\tilde{G}(u,v) = \frac{1}{2}H(u)H(v). \tag{5.66}$$

It is one half inside the future light cone and zero outside. A point-like perturbation propagates with all velocities between plus c and minus c. Note that this Green function not only vanishes for negative times, but also for space-like points (t,x), i.e. $t^2 - x^2 < 0$. This means that the Green function is not only causal in our inertial frame, it is causal in any inertial frame.

In the old coordinates the Green function is

$$G(t,x) = \frac{1}{2}H(t)[H(x+t) - H(x-t)]. \tag{5.67}$$

Inserting into equation (5.57) we obtain the explicit solution of the homogeneous Cauchy problem, $s \equiv 0$:

$$f(t,x) = \frac{1}{2}H(t)\left\{\int [H(x-\xi+t) - H(x-\xi-t)]\,h(\xi)d\xi\right.$$

$$\left. + \int [\delta(x-\xi+t) + \delta(x-\xi-t)]g(\xi)d\xi\right\} \tag{5.68}$$

$$= H(t)\left[\frac{1}{2}\int_{x-t}^{x+t} h(\xi)d\xi + \frac{1}{2}[g(x-t) + g(x+t)]\right].$$

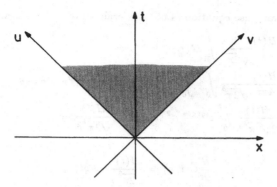

Fig. 5.2   The causal, Lorentz-invariant Green function

In agreement with equation (5.65) the propagation velocity is $\pm c$ if there are no initial velocities, i.e. $h(x) \equiv 0$.

$\underline{N = 2}$

The two-dimensional Green function can be obtained from the three-dimensional one by the so-called descendent method. Details can be found in Schwartz (1965), for example. One gets:

$$G(t, x) = \frac{H(t)}{2\pi} \frac{H(t - r)}{\sqrt{t^2 - r^2}}. \tag{5.69}$$

The support of G is again the future light cone and its interior. However, this time propagation velocities close to c are favoured by the singularity on the light cone.

$\underline{N = 3}$

Here, in three dimensions, we encounter our first Green function that is not a function:

$$\begin{aligned} G(t, x) &= \frac{H(t)}{2\pi} \delta(r^2 - t^2) \\ &= \frac{H(t)}{4\pi t} [\delta(r + t) + \delta(r - t)] \\ &= \frac{H(t)}{4\pi t} \delta(r - t) \end{aligned} \tag{5.70}$$

where we have used equation (2.38). Lorentz invariance of the causal Green function is manifest in the first form given. To prove that $G(t, x)$ is a Green function, we again Fourier-transform with respect to x only. In the spatial variables G is of compact support and we may use equation (4.65) (generalized to three dimensions):

$$
\begin{aligned}
\hat{G}(t,k) &= \frac{H(t)}{4\pi t}\frac{1}{\sqrt{2\pi}^3}\int_{\mathbb{R}^3}\delta(r-t)e^{i k\cdot x}dx \\
&= \frac{H(t)}{4\pi t}\frac{1}{\sqrt{2\pi}^3}\int_0^\infty\int_0^{2\pi}\int_0^\pi\delta(r-t)e^{iqr\cos\theta}r^2\sin\theta d\theta d\phi dr \qquad (5.71) \\
&= \frac{H(t)}{2\sqrt{2\pi}^3 t}\int_0^\pi e^{iqt\cos\theta}t^2\sin\theta d\theta = \frac{H(t)}{\sqrt{2\pi}^3}\frac{\sin qt}{q}.
\end{aligned}
$$

Then

$$
\frac{\partial}{\partial t}\hat{G}(t,k) = \frac{H(t)}{\sqrt{2\pi}^3}\frac{\cos qt}{q}, \qquad (5.72)
$$

$$
\frac{\partial^2}{\partial t^2}\hat{G}(t,k) = \frac{1}{\sqrt{2\pi}^3}\delta(t) - \frac{H(t)}{\sqrt{2\pi}^3}q\sin qt. \qquad (5.73)
$$

Finally

$$
\left(\frac{\partial^2}{\partial t^2}+q^2\right)\hat{G}(t,k) = \delta(t)\frac{1}{\sqrt{2\pi}^3} = \delta(t)\hat{\delta}(k). \qquad (5.74)
$$

In three dimensions the support of the causal Green function is the future light cone without the interior, signals propagate only at the speed of light. The causal Green function is often referred to as retarded Green function in contrast to the advanced Green function

$$
G^{adv}(t,x) = \frac{H(-t)}{2\pi}\delta(r^2-t^2) = \frac{-H(-t)}{2\pi}\delta(r+t). \qquad (5.75)
$$

The causal Green function describes a transmitting antenna where a forced electric current j^μ in the antenna is the source of electromagnetic radiation given by the potential

$$
A^\mu(ct,c) = \frac{1}{4\pi\epsilon_0 c^2}\int_{\mathbb{R}^4} G(ct-c\tau, x-\xi)j^\mu(c\tau,\xi)d(c\tau)d\xi. \qquad (5.76)
$$

In particular, if j^μ is the current of an accelerated point charge, A^μ are the well-known Liénard-Wiechert potentials. On the other hand, the advanced Green function in equation (5.76) relates an induced electric current in a receiving antenna with the incoming radiation.

For completeness we indicate these two Green functions using the complex representation (5.44) of the step function $H(t)$:

$$
\hat{G}(\omega,k) = \frac{1}{(2\pi)^2}\frac{1}{-\omega^2+q^2-i\omega\epsilon}, \qquad (5.77)
$$

$$
\hat{G}^{adv}(\omega,k) = \frac{1}{(2\pi)^2}\frac{1}{-\omega^2+q^2+i\omega\epsilon}. \qquad (5.78)
$$

Without the ϵ they would both be equal to $(2\pi)^{-2} P(\omega, k)^{-1}$ explaining why sometimes the Green function is addressed as an inverse of the differential operator, $G = \square^{-1}$.

We end this section with a remark about the support of the causal Green function in higher dimensions. It is the future light cone in odd dimensions $N = 3, 5...$ and the future light cone together with its interior in even dimensions $N = 2, 4...$ For more details the reader is referred to Choquet-Bruhat & DeWitt-Morette (1989).

Exercises

1. Consider the subset $\{(t, x) \in \mathbb{R}^4, \ t > 0\}$ of space time. What is its image under all Lorentz boosts?

2. Verify directly in the coordinates t and x that the Green function (5.67) satisfies $\square G = \delta$.

3. Check that

$$f(x, t) = \frac{1}{2} \int_{x-t}^{x+t} h(\xi) d\xi + \frac{1}{2}[g(x - t) + g(x + t)]$$

solves the homogeneous Cauchy problem of the wave equation in 1+1 dimensions.

4. In 1+1 dimensions solve the inhomogeneous Cauchy problem

$$\square f(t, x) = \sin(at - bx), \quad a, b \in \mathbb{R},$$
$$f(0, x) = \frac{\partial}{\partial t} f(0, x) = 0.$$

5.4 The Klein-Gordon equation

"We see a world in a mirror.
Parallel to ours, precisely reversed.
A world with its own rules,
yet dependent on ours, as we on it."
Glass, Hwang & Sirlin (1989).

In 1924 de Broglie gave precise meaning to the particle-wave duality by introducing relations between momentum and wave vector and between total energy and angular frequency:

$$p = \hbar k, \tag{5.79}$$

$$E = \hbar \omega. \tag{5.80}$$

In these variables the polynomial of the free Schrödinger equation becomes

$$P(\omega, k) = -\frac{2m}{\hbar}\omega + k^2 = \frac{-2m}{\hbar^2}\left[E - \frac{p^2}{2m}\right]. \tag{5.81}$$

Turning around this argument we can start from energy conservation in Newtonian mechanics,

$$E - \frac{p^2}{2m} = 0, \tag{5.82}$$

to arrive at the Schrödinger equation of a particle of mass m and spin zero:

$$\left(E - \frac{p^2}{2m}\right)\hat{\psi}(E, p) = 0. \tag{5.83}$$

Likewise energy conservation of relativistic mechanics

$$E^2 - c^2 p^2 - c^4 m^2 = 0 \tag{5.84}$$

yields

$$(\hbar^2 \omega^2 - c^2 \hbar^2 k^2 - c^4 m^2)\hat{\psi} = 0 \tag{5.85}$$

and

$$\left(\frac{1}{c^2}\frac{\partial^2}{\partial t^2} - \Delta + \frac{c^2}{\hbar^2}m^2\right)\psi = \left(\Box + \frac{c^2}{\hbar^2}m^2\right)\psi = 0. \tag{5.86}$$

This is the free Klein-Gordon equation. It describes a massive relativistic particle with spin zero. For simplicity let us put $\hbar = c = 1$. The retarded Green function of the Klein-Gordon operator $\Box + m^2$ in $N = 3$ dimensions is

$$G(t, x) = \frac{H(t)}{2\pi}[\delta(r^2 - t^2) - H(t^2 - r^2)\frac{m}{2\sqrt{t^2 - r^2}}J_1(m\sqrt{t^2 - r^2})] \tag{5.87}$$

where J_1 is the first-order Bessel function to be defined in chapter 7. In the massless case, of course, we retrieve the propagator of the wave equation (5.70). For positive masses, however, the support extends into the interior of the future light cone, due to the Bessel function. In the ϵ notation retarded and advanced Green functions take a simple form:

$$\hat{G}(\omega, k) = \frac{1}{(2\pi)^2} \frac{1}{m^2 - \omega^2 + q^2 - i\omega\epsilon} \tag{5.88}$$

$$\hat{G}^{adv}(\omega, k) = \frac{1}{(2\pi)^2} \frac{1}{m^2 - \omega^2 + q^2 + i\omega\epsilon}. \tag{5.89}$$

The Klein-Gordon equation presents an apparent difficulty: Unlike with the Schrödinger equation the total energy is not positive, $E = \pm\sqrt{m^2 + p^2}$. In fact Schrödinger had studied the Klein-Gordon equation already in 1926 before he considered his non-relativistic equation to which he went on precisely to avoid the difficulties connected with negative energies. These difficulties where turned into brilliant success by Dirac. In 1930 he proposed to interpret a particle of negative energy as antiparticle thereby postulating antimatter for every existing form of matter. First experimental verification of his daring prediction came in 1932 with the discovery of the positron by Anderson. The possibility of producing a pair of particle-antiparticle now prohibits any ordinary interpretation of the Klein-Gordon wave function ψ and forces upon us a theoretical frame work that goes far beyond quantum mechanics. This frame work is quantum field theory. One of its basic ingredients is the hypothesis due to Stückelberg (1941) and Feynman (1949) that antiparticles propagate backwards in time. They introduced a Green function, today called Feynman propagator, that is retarded for positive energies but advanced for negative energies:

$$\hat{G}^F(\omega, k) = \frac{1}{(2\pi)^2} \frac{1}{m^2 - \omega^2 + q^2 - i\epsilon}. \tag{5.90}$$

For curiosity we also indicate the Feynman propagator in space time.

$$\begin{aligned} G^F(t, x) = &\frac{1}{4\pi}[\delta(r^2 - t^2) - H(t^2 - r^2)\frac{m}{2\sqrt{t^2 - r^2}} J_1(m\sqrt{t^2 - r^2})] \\ &+ \frac{i}{4\pi} H(t^2 - r^2)\frac{m}{2\sqrt{t^2 - r^2}} N_1(m\sqrt{t^2 - r^2}) \\ &- \frac{i}{4\pi} H(r^2 - t^2)\frac{m}{\pi\sqrt{r^2 - t^2}} K_1(m\sqrt{r^2 - t^2}) \end{aligned} \tag{5.91}$$

where N_1 is the Neumann function of first order, K_1 the modified Bessel function of first order. Due to the last term, the support of the Feynman propagator extends not only over the light cone but also into space-like regions. Nevertheless the Feynman propagator is also called causal propagator in quantum field theory.

All three propagators $G(t, x)$, the retarded, the advanced, and the Feynman one, are obtained from their "Fourier transforms" $\hat{G}_\epsilon(\omega, k)$ by

$$G(t, x) = \lim_{\epsilon \to 0+} \int_{\mathbb{R}^3} \int_{-\infty}^{+\infty} \hat{G}_\epsilon(\omega, k)e^{-i(\omega t - k \cdot x)} d\omega dk. \tag{5.92}$$

112

The positive ϵ in $\hat{G}_\epsilon(\omega, k)$ fixes the character of the Green function by specifying how the integration curve, the real ω-axis, avoids the singularities of the integrand. E.g. in the retarded case the singularities are positioned at

$$\omega = -\frac{\mathrm{i}}{2}\epsilon \pm \sqrt{m^2 + q^2 - \frac{\epsilon^2}{4}}. \tag{5.93}$$

By Cauchy's theorem we can equivalently keep the singularities of the integrand on the real ω-axis and perform the ω integration over a complex curve that avoids the real singularities by ϵ:

$$G(t, x) = \lim_{\epsilon \to 0+} \int_{\mathbb{R}^3} \int_C \frac{1}{(2\pi)^2} \frac{e^{-\mathrm{i}(\omega t - k \cdot x)}}{m^2 - \omega^2 + q^2} d\omega \, dk \tag{5.94}$$

where C is one of the three curves shown in fig. 5.3.

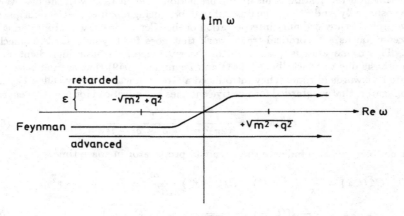

Fig. 5.3   Three integration curves to obtain retarded, advanced, and Feynman propagators

Exercises

1. For what energies ω_0 is the plane wave

$$\psi(t, x) = e^{-\mathrm{i}(\omega_0 t - k_0 \cdot x)}$$

solution of the homogeneous Klein-Gordon equation? What is its Fourier transform?

2. Explain why the real Klein-Gordon equation has the complex Green function (5.91).

5.5 The Dirac equation

Dirac was looking for a Lorentz invariant generalization of the Schrödinger equation that remained first order in time. He found that this was possible only for a system of wave functions, called spinor,

$$\psi(x^0, x) = \begin{pmatrix} \psi_1(x^0, x) \\ \psi_2(x^0, x) \\ \psi_3(x^0, x) \\ \psi_4(x^0, x) \end{pmatrix} \tag{5.95}$$

and a system of differential equations.

$$\left(i\hbar \sum_{\mu=0}^{3} \gamma^\mu \frac{\partial}{\partial x^\mu} - cm\mathbb{1}_4 \right) \psi(x^0, x) = 0. \tag{5.96}$$

Here γ^μ, $\mu = 0, 1, 2, 3$ are four complex 4×4 matrices called Dirac matrices:

$$\gamma^0 := \begin{pmatrix} \mathbb{1}_2 & 0 \\ 0 & \mathbb{1}_2 \end{pmatrix} \tag{5.97}$$

$$\gamma^j := \begin{pmatrix} 0 & \sigma^j \\ \sigma^j & 0 \end{pmatrix}, \quad j = 1, 2, 3 \tag{5.98}$$

with the Pauli matrices

$$\sigma^1 := \begin{pmatrix} 0 & 1 \\ 1 & 0 \end{pmatrix}, \tag{5.99}$$

$$\sigma^2 := \begin{pmatrix} 0 & -i \\ i & 0 \end{pmatrix}, \tag{5.100}$$

$$\sigma^3 := \begin{pmatrix} 1 & 0 \\ 0 & -1 \end{pmatrix}. \tag{5.101}$$

We denote the 2×2 and 4×4 unit matrix by $\mathbb{1}_2$ and $\mathbb{1}_4$. Often the unit matrix in the Dirac equation (5.96) is omitted. Note that the spinor index $A = 1, 2, 3, 4$ should not be confused with the space-time index $\mu = 0, 1, 2, 3$. Two components of the spinor ψ describe a spin $\frac{1}{2}$ particle of mass m, e.g. the electron, the other two components describe the corresponding antiparticle with the same mass and spin, e.g. the positron.

The Dirac operator $\partial\!\!\!/$ is by definition the 4×4 matrix of linear differential operators with constant coefficients

$$\partial\!\!\!/ := \sum_{\mu=0}^{3} \gamma^\mu \frac{\partial}{\partial x^\mu}. \tag{5.102}$$

Its square is the d'Alembert operator:

$$\partial\!\!\!/^2 = \mathbb{1}_4 \square. \tag{5.103}$$

The proof of this equation makes use of the following anticommutation relations for the Dirac matrices:

$$\gamma^\mu \gamma^\nu + \gamma^\nu \gamma^\mu = 2\eta^{\mu\nu} \mathbb{1}_4. \tag{5.104}$$

Setting again $\hbar = c = 1$, the Dirac equation becomes

$$(i\partial\!\!\!/ - m)\psi = 0 \tag{5.105}$$

and every component of ψ solving the homogeneous Dirac equation also satisfies the homogeneous Klein-Gordon equation:

$$\begin{aligned}
0 &= (i\partial\!\!\!/ + m)(i\partial\!\!\!/ - m)\psi = (-\partial\!\!\!/^2 - m^2)\psi \\
&= -(\square + m^2)\psi.
\end{aligned} \tag{5.106}$$

A Green function of $(i\partial\!\!\!/ - m)$ is a 4×4 matrix of distributions denoted traditionally by $S(x^0, x)$ such that

$$(i\partial\!\!\!/ - m)S = \delta(x^0)\delta(x)\mathbb{1}_4. \tag{5.107}$$

The retarded, advanced and Feynman propagators $S(x^0, x)$ are obtained easily from the corresponding propagator $G(x^0, x)$ of the Klein-Gordon operator:

$$S = -(i\partial\!\!\!/ + m)G. \tag{5.108}$$

In Fourier space for instance we find

$$\hat{S}^F(k^0, k) = \frac{1}{(2\pi)^2} \frac{-\sum_{\mu=0}^{3} \gamma^\mu \eta_{\mu\nu} k^\nu - m}{m^2 - \omega^2 + q^2 - i\epsilon} \tag{5.109}$$

where the Fourier transform is understood with respect to the Minkowskian scalar product $k^0 x^0 - k \cdot x$.

To close this chapter let us remark that the Laplace, d'Alembert, and Dirac operators can be generalized from flat $\mathbb{R}^N$ to curved manifolds. For instance the Laplace operator on the two-sphere of radius r is obtained from the Laplace operator in $\mathbb{R}^3$ in polar coordinates (Exercise 2, section 5.1) by deleting partial derivatives with respect to r. However, on manifolds the techniques outlined in this chapter fail for several reasons. There are no tempered distributions and no Fourier transform on arbitrary manifolds. The Laplace, d'Alembert, and Dirac operators although still linear have non-constant coefficients on curved manifolds, as illustrated by the Laplace operator on the two-sphere. For these reasons also any attempt to define quantum field theory on curved space time has failed so far. In chapter 7 we shall learn a few generalizations of the Fourier techniques to highly symmetric manifolds like sphere and cylinder.

Exercises

1. Verify the anticommutation relations (5.104) of the Dirac matrices.
2. Show that the Dirac operator is a "square root" of the d'Alembert operator, i.e. verify equation (5.103).
3. Derive equation (5.109).

6

Linear algebra in infinite dimensions

The purpose of this chapter is to show that the material covered so far naturally fits into the realm of linear algebra in infinite dimensions. This point of view will suggest very fruitful generalizations of the Fourier series.

6.1 Complex vector spaces

To begin, let us recall the definition of a vector space.

Definition: A (complex) vector space V is a set equipped with two operations: addition

$$V \times V \to V$$

$$(v, w) \to v + w$$

and scalar multiplication

$$\mathbb{C} \times V \to V$$

$$(a, v) \to av$$

satisfying the following axioms:
1. Commutativity

$$v + w = w + v. \tag{6.1}$$

2. Associativity

$$(v + w) + u = v + (w + u). \tag{6.2}$$

3. There is a neutral element $0 \in V$ such that for all vectors $v \in V$

$$v + 0 = v. \tag{6.3}$$

4. Every vector $v \in V$ has an inverse denoted by $-v$ satisfying

$$v + (-v) = 0. \tag{6.4}$$

5. For every vector v

$$1v = v. \tag{6.5}$$

6. For all scalars $a, b \in \mathbb{C}$ and every vector v

$$(ab)v = a(bv). \tag{6.6}$$

116

7. Distributivity

$$a(v + w) = av + aw, \tag{6.7}$$

$$(a + b)v = av + bv. \tag{6.8}$$

The important finite dimensional example is $V = \mathbb{C}^N$. Let us denote its elements by column vectors

$$v = \begin{pmatrix} v^1 \\ \vdots \\ v^N \end{pmatrix}, \qquad v^j \in \mathbb{C}. \tag{6.9}$$

Addition and scalar multiplication are defined component-wise:

$$v + w := \begin{pmatrix} v^1 + w^1 \\ \vdots \\ v^N + w^N \end{pmatrix}, \tag{6.10}$$

$$av := \begin{pmatrix} av^1 \\ \vdots \\ av^N \end{pmatrix}. \tag{6.11}$$

The neutral element is the zero vector

$$0 := \begin{pmatrix} 0 \\ \vdots \\ 0 \end{pmatrix}. \tag{6.12}$$

Every finite dimensional vector space is isomorphic to $\mathbb{C}^N$ for some N. The infinite dimensional vector spaces are more interesting. There is for instance the space of all polynomials p in one real variable x and with complex coefficients v^j

$$p(x) = v^0 + v^1 x + v^2 x^2 + \ldots + v^m x^m, \quad m \in \mathbb{N}_0 \tag{6.13}$$

This vector space has as N dimensional subspace the set of all polynomials of degree less than or equal to $N - 1$. Recall that a subspace of a vector space is a subset that itself is a vector space under the same addition and scalar multiplication.

We have already made use of another vector space of infinite dimension, the set of all (infinitely many times) differentiable functions of several real variables and with complex values. Let Ω, the domain of definition, be a convenient subset of $\mathbb{R}^N$, e.g. the entire $\mathbb{R}^N$, a hypercube or the circle. Then this space of functions is denoted by $\mathcal{C}^\infty(\Omega)$. The test functions form a subspace which we call $\mathcal{C}_0^\infty(\Omega)$. As we have seen in chapter 2, also distributions may be added and scalar-multiplied. Therefore they form an even bigger vector space denoted by $\mathcal{D}(\Omega)$ with nested subspaces:

$$\mathcal{C}_0^\infty(\Omega) \subset \mathcal{C}^\infty(\Omega) \subset \mathcal{D}(\Omega). \tag{6.14}$$

Exercises

1. Prove that the continuous square integrable functions from $\mathbb{R}^N$ into $\mathbb{C}$ form a complex vector space. Hint: Use the Cauchy-Schwarz inequality.
2. Is the set of all invertible complex functions a vector space?
3. Show that the tempered distributions and the functions of fast decrease are two subspaces of $\mathcal{D}(\mathbb{R}^N)$. Complete the inclusion relations (6.14) with these subspaces.

6.2 Operators

Definition: An operator (or linear transformation) A on a vector space V is a linear mapping from V into itself:

$$A(v + w) = Av + Aw, \tag{6.15}$$

$$A(av) = aAv \tag{6.16}$$

for all vectors $v, w \in V$ and scalars $a \in \mathbb{C}$.

For the finite dimensional vector space $V = \mathbb{C}^N$ an operator A is simply a $N \times N$ matrix with complex entries $A^j{}_k$:

$$A \begin{pmatrix} v^1 \\ \vdots \\ v^N \end{pmatrix} = \begin{pmatrix} A^1{}_1 v^1 + A^1{}_2 v^2 + \dots + A^1{}_N v^N \\ \vdots \\ A^N{}_1 v^1 + A^N{}_2 v^2 + \dots + A^N{}_N v^N \end{pmatrix}. \tag{6.17}$$

On infinite dimensional spaces there are more interesting operators, for example

- partial differentiation

$$\frac{\partial}{\partial x^k} : \mathcal{C}^\infty(\mathbb{R}^N \to \mathcal{C}^\infty(\mathbb{R}^N), \quad 1 \le k \le N$$

$$f(x) \mapsto \frac{\partial f}{\partial x^k}(x) \tag{6.18}$$

- multiplication with a fixed differentiable function $h(x)$

$$h \cdot : \mathcal{C}^\infty(\mathbb{R}^N) \to \mathcal{C}^\infty(\mathbb{R}^N)$$

$$f(x) \mapsto h(x) f(x) \tag{6.19}$$

- any linear differential operator, not necessarily with constant coefficients, like

$$A = -\frac{d^2}{dx^2} + x^2 \tag{6.20}$$

- the convolution with a fixed test function $h(x)$

$$h* : \mathcal{C}^{\infty}(\mathbb{R}^N) \to \mathcal{C}^{\infty}(\mathbb{R}^N)$$
$$f(x) \mapsto (h * f)(x) \tag{6.21}$$

- more generally let $K(x, \xi)$ be a test function on $\mathbb{R}^{2N}$ and define the integral operator with kernel K by

$$A : \mathcal{C}^{\infty}(\mathbb{R}^N) \to \mathcal{C}^{\infty}(\mathbb{R}^N)$$
$$f(x) \mapsto \int_{\mathbb{R}^N} K(x, \xi) f(\xi) d\xi \tag{6.22}$$

- the Fourier transform

$$\hat{} : \mathcal{S}(\mathbb{R}^N) \to \mathcal{S}(\mathbb{R}^N)$$
$$f(x) \mapsto \hat{f}(k) \tag{6.23}$$

where $\mathcal{S}(\mathbb{R}^N)$ is the space of functions of fast decrease.
- integration with respect to $x^k, 1 \le k \le N$

$$\int_0^{x^k} : \mathcal{C}^{\infty}(\mathbb{R}^N) \to \mathcal{C}^{\infty}(\mathbb{R}^N)$$

$$f(x^1, ..., x^N) \mapsto \int_0^{x^k} f(x^1, ..., x^{k-1}, \xi, x^{k+1}, ..., x^N) d\xi. \tag{6.24}$$

The sum of two operators A and B defined as usual by

$$(A + B)(v) := Av + Bv \tag{6.25}$$

is again an operator as for example in (6.20). Similarly any scalar multiple of an operator

$$(aA)(v) := aAv \tag{6.26}$$

is an operator and the operators on a given vector space from themselves a vector space. Furthermore the composition of two operators A and B

$$(A \circ B)(v) := A(Bv) \tag{6.27}$$

is again an operator, e.g. $d^2/dx^2 = d/dx \circ d/dx$ and the vector space of operators becomes an algebra with composition as product. This product is associative, non-commutative, in general, and it admits as neutral element the identity mapping $\mathbb{1}$ on V,

$$\mathbb{1}v = v \tag{6.28}$$

for all vectors $v \in V$. For simplicity, we omit the composition sign in the product of operators from now on. For $V = \mathbb{C}^N$ this product is simply the product of $N \times N$ matrices. In general we denote by $[A, B]$ the commutator of two operators:

$$[A, B] := AB - BA. \tag{6.29}$$

An operator A is said to be invertible if there is a second operator denoted by A^{-1} such that

$$AA^{-1} = A^{-1}A = \mathbb{1}. \tag{6.30}$$

To show that an operator A on a finite dimensional vector space is invertible, it is sufficient to exhibit a right inverse A^{-1} satisfying

$$AA^{-1} = \mathbb{1}. \tag{6.31}$$

The right inverse then automatically is a left inverse

$$A^{-1}A = \mathbb{1}. \tag{6.32}$$

This is not true for operators on infinite dimensional spaces: For instance d/dx on $\mathcal{C}^{\infty}(\mathbb{R})$ has a right inverse $\int_0^x$ but no left inverse.

A useful tool to decide if an operator is invertible are its eigenvalues.

Definition: Let A be an operator on the vector space V. A complex number λ is called eigenvalue of A if there is a non-zero vector $v \in V$, such that

$$Av = \lambda v. \tag{6.33}$$

In this case v is called eigenvector to the eigenvalue λ and the subset of $\mathbb{C}$ consisting of all eigenvalues is called the point spectrum of A.

Examples: Consider $V = \mathbb{C}^2$ and

$$A = \begin{pmatrix} i & 1 \\ 0 & \sqrt{2} \end{pmatrix}. \tag{6.34}$$

Then i is eigenvalue with eigenvector $\begin{pmatrix} 1 \\ 0 \end{pmatrix}$ because

$$\begin{pmatrix} i & 1 \\ 0 & \sqrt{2} \end{pmatrix} \begin{pmatrix} 1 \\ 0 \end{pmatrix} = \begin{pmatrix} i \\ 0 \end{pmatrix} = i \begin{pmatrix} 1 \\ 0 \end{pmatrix}. \tag{6.35}$$

Every operator on a finite dimensional complex vector space has at least one eigenvalue. This is not true over real vector spaces, e.g.

$$A = \begin{pmatrix} 0 & -1 \\ 1 & 0 \end{pmatrix} \tag{6.36}$$

has no real eigenvalue.

Consider $V = \mathcal{C}^{\infty}(\mathbb{R})$ and $A = d/dx$. Here any $\lambda \in \mathbb{C}$ is eigenvalue with eigenvector $e^{\lambda x}$:

$$\frac{d}{dx} e^{\lambda x} = \lambda e^{\lambda x}. \tag{6.37}$$

Since any operator A maps the zero vector into itself

$$A(0) = 0 \qquad (6.38)$$

operators with a vanishing eigenvalue cannot be invertible, as d/dx, for example.

Consider the multiplication by x on the space of distributions $\mathcal{D}(\mathbb{R})$. Every real number λ is eigenvalue with eigenvector $\delta(x - \lambda)$:

$$x\delta(x - \lambda) = \lambda\delta(x - \lambda). \qquad (6.39)$$

Exercises

1. Which of the operators (6.18-24) can be restricted to $\mathcal{C}_0^\infty(\mathbb{R}^N)$?
2. Which of the operators (6.18-24) can be extended to $\mathcal{D}(\mathbb{R}^N)$?
3. If possible, indicate the inverse of the following operators:
   a) d/dx on $\mathcal{C}_0^\infty(\mathbb{R})$,
   b) multiplication by x on $\mathcal{C}^\infty(\mathbb{R})$,
   c) multiplication by $x^2 + 1$ on $\mathcal{D}(\mathbb{R})$.
4. Does d/dx have a vanishing eigenvalue as operator on $\mathcal{D}(\mathbb{R}), \mathcal{C}^\infty(\mathbb{R}), \mathcal{S}(\mathbb{R}), \mathcal{C}_0^\infty(\mathbb{R})$?
5. What is the point spectrum of the following operators:
   a) $\begin{pmatrix} 0 & -1 \\ 1 & 0 \end{pmatrix}$ on $\mathbb{C}^2$,
   b) d/dx on $\mathcal{D}(\mathbb{R}), \mathcal{T}(\mathbb{R}), \mathcal{C}^\infty(\mathbb{R}), \mathcal{S}(\mathbb{R})$, where $\mathcal{T}(\mathbb{R})$ is the space of tempered distributions, and on $\mathcal{C}^\infty(\text{circle})$,
   c) multiplication by x on $\mathcal{C}^\infty(\mathbb{R})$,
   d) multiplication by $x^2 + 1$ on $\mathcal{D}(\mathbb{R})$?

6.3 Scalar products

Definition: Let V be a complex vector space. A scalar product (or hermitian metric) on V is a mapping

$$V \times V \rightarrow \mathbb{C}$$

$$(v, w) \rightarrow (v, w)$$

with the following properties:
   1. linearity in the second variable:

$$(v, w_1 + w_2) = (v, w_1) + (v, w_2), \qquad (6.40)$$

$$(v, aw) = a(v, w), \qquad (6.41)$$

   2. Hermiticity:

$$(w, v) = \overline{(v, w)}, \qquad (6.42)$$

3. positivity:

$$(v, v) > 0 \tag{6.43}$$

for all non-zero vectors $v \in V$.

Note that in most of the mathematical literature linearity is postulated in the first variable, not in the second.

Examples: On $V = \mathbb{C}^N$ we define the so-called standard scalar product:

$$(v, w) := \sum_{j=1}^{N} \bar{v}^j w^j. \tag{6.44}$$

By the Gram-Schmidt orthonormalization procedure any scalar product on a finite dimensional vector space is equivalent to the standard scalar product.

On the vector space of all square summable sequences of complex numbers:

$$\ell^2 := \{ v = (v^1, v^2, v^3, \ldots), v^j \in \mathbb{C}, \sum_{j=1}^{\infty} |v^j|^2 < \infty \} \tag{6.45}$$

we can define a scalar product that generalizes the above one:

$$(v, w) := \sum_{j=1}^{\infty} \bar{v}^j w^j. \tag{6.46}$$

On the vector space of test functions $\mathcal{C}_0^\infty(\Omega)$ we define the scalar product:

$$(f, g) := \int_\Omega \bar{f} g. \tag{6.47}$$

A few immediate properties of any scalar product are:

1. Antilinearity in the first variable:

$$(v_1 + v_2, w) = (v_1, w) + (v_2, w), \tag{6.48}$$

$$(av, w) = \bar{a}(v, w). \tag{6.49}$$

In words, every complex number a, when pulled out of the first entry picks up a complex conjugate.

2. The scalar product is non-degenerate, i.e. only the zero vector can have vanishing scalar product with every vector in V.

3. Hermiticity implies that the scalar product of a vector with itself is a real number, which is non-negative by the positivity axiom. Therefore we can define the norm (or length) of a vector v by

$$|v| := \sqrt{(v, v)}. \tag{6.50}$$

and a vector has vanishing norm if and only if the vector is zero.

4. The Cauchy-Schwarz inequality:

$$|(v, w)| \leq |v||w|. \tag{6.51}$$

Proof: If the scalar product (v, w) vanishes, the inequality holds trivially. Otherwise

$$
\begin{aligned}
0 \leq & \left| \frac{v}{|v|} - \frac{\overline{(v, w)}}{|(v, w)|} \frac{w}{|w|} \right| = \left(\frac{v}{|v|} - \frac{\overline{(v, w)}}{|(v, w)|} \frac{w}{|w|}, \frac{v}{|v|} - \frac{\overline{(v, w)}}{|(v, w)|} \frac{w}{|w|} \right) \\
= & \frac{(v, v)}{|v|^2} + \frac{\overline{(v, w)}}{|(v, w)|} \frac{(v, w)}{|(v, w)|} \frac{(w, w)}{|w|^2} - \frac{\overline{\overline{(v, w)}}}{|(v, w)|} \frac{(w, v)}{|v||w|} - \frac{\overline{(v, w)}}{|(v, w)|} \frac{(v, w)}{|v||w|} \quad (6.52) \\
= & 2 - 2 \frac{|(v, w)|^2}{|(v, w)||v||w|} = 2 \left(1 - \frac{|(v, w)|}{|v||w|} \right).
\end{aligned}
$$

5. Triangle inequality:

$$|v + w| \leq |v| + |w|. \tag{6.53}$$

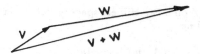

Fig. 6.1   The triangle inequality

The proof uses the triangle inequality of complex numbers:

$$
\begin{aligned}
|v + w|^2 &= (v + w, v + w) = |v|^2 + |w|^2 + 2\mathrm{Re}(v, w) \\
&\leq |v|^2 + |w|^2 + 2|(v, w)| \leq |v|^2 + |w|^2 + 2|v||w| = (|v| + |w|)^2.
\end{aligned} \tag{6.54}
$$

A vector space together with a scalar product is called a Pre-Hilbert space. There are two most important classes of operators on a Pre-Hilbert space: the Hermitian operators and the unitary operators.

Definition: An operator U on a Pre-Hilbert space V is called unitary if it is invertible and if it preserves the scalar product:

$$(Uv, Uw) = (v, w) \tag{6.55}$$

for all vectors $v, w \in V$.

For $V = \mathbb{C}^N$ with the standard scalar product, a matrix U is a unitary operator if and only if its inverse equals its complex conjugate, transposed matrix,

$$U^{-1} = \bar{U}^T. \tag{6.56}$$

By Parseval's theorem (4.29) the Fourier transform on $\mathcal{S}(\mathbb{R}^N)$, the space of functions of fast decrease, is a unitary operator with respect to the scalar product

$$(f,g) := \int_{\mathbb{R}^N} \bar{f}g \tag{6.57}$$

of two functions f and g of fast decrease. Other obviously unitary operators on $\mathcal{S}(\mathbb{R}^N)$ are induced by translations with a constant vector $\xi \in \mathbb{R}^N$ and by rotations $R \in O(N)$:

$$(U_\xi f)(x) := f(x - \xi), \tag{6.58}$$

$$(U_R f)(x) := f(R^{-1}x). \tag{6.59}$$

The point spectrum of any unitary operator is contained in the unit circle of the complex plane. Indeed if λ is an eigenvalue with eigenvector v

$$Uv = \lambda v \tag{6.60}$$

then

$$(Uv, Uv) = (\lambda v, \lambda v) = |\lambda|^2 |v|^2 = |v|^2. \tag{6.61}$$

Definition: An operator A on a Pre-Hilbert space is called Hermitian (or symmetric) if for all vectors v and w:

$$(Av, w) = (v, Aw). \tag{6.62}$$

For $V = \mathbb{C}^N$ with the standard scalar product a matrix A is a Hermitian operator if and only if it equals its complex conjugate transposed:

$$A = \bar{A}^T. \tag{6.63}$$

The multiplication operator $x^k \cdot$ and i times the partial differentiation $i\partial/\partial x^k$ are both Hermitian operators on $C_0^\infty(\mathbb{R}^N)$ with respect to the scalar product (6.57). Likewise id/dt on $C_0^\infty(\text{circle})$ is Hermitian.

The point spectrum of any Hermitian operator is real. Indeed if λ is an eigenvalue with eigenvector v

$$Av = \lambda v \tag{6.64}$$

then

$$(Av, v) = (\lambda v, v) = \bar{\lambda}|v|^2 = (v, Av) = (v, \lambda v) = \lambda|v|^2. \tag{6.65}$$

Furthermore two eigenvectors v and w with different eigenvalues $\lambda \neq \mu$ of the Hermitian operator A,

$$Aw = \mu w \tag{6.66}$$

are orthogonal, i.e.

$$(v, w) = 0. \tag{6.67}$$

Indeed

$$0 = (Av, w) - (v, Aw) = (\lambda - \mu)(v, w). \qquad (6.68)$$

For instance the harmonics

$$e_j(t) = \frac{1}{\sqrt{T}} e^{-ij2\pi t/T} \qquad (6.69)$$

are eigenvectors of the Hermitian operator id/dt on C_0^∞(circle) with distinct real eigenvalues

$$\lambda_j = j\frac{2\pi}{T}. \qquad (6.70)$$

Therefore they are all orthogonal.

We are ready now for the proof of the uncertainty relation of Fourier transforms. Let A be a Hermitian operator and $v \in V$ a non-zero vector. We define the expectation of A with respect to v (or the mean value of A in the state v) by

$$\langle A \rangle := \frac{(v, Av)}{(v, v)}. \qquad (6.71)$$

Since A is Hermitian, its expectation is always real. We define the standard deviation of A with respect to v (or the uncertainty of A in the state v) by

$$\Delta A := \sqrt{\langle (A - \langle A \rangle \mathbb{1})^2 \rangle} = \sqrt{\langle A^2 \rangle - \langle A \rangle^2}. \qquad (6.72)$$

If A and B are two Hermitian operators, then $i[A, B]$ is again Hermitian and we have the uncertainty relation

$$(\Delta A)(\Delta B) \geq \frac{1}{2} |\langle i[A, B] \rangle|. \qquad (6.73)$$

Proof: For convenience we set

$$\tilde{A} := A - \langle A \rangle \mathbb{1}, \qquad (6.74)$$

$$\tilde{B} := B - \langle B \rangle \mathbb{1}. \qquad (6.75)$$

It follows that

$$[\tilde{A}, \tilde{B}] = [A, B]. \qquad (6.76)$$

For any real number c we have the inequality

$$\begin{aligned} 0 &\leq |(\tilde{A} + ic\tilde{B})v|^2. \\ &= |\tilde{A}v|^2 + c^2|\tilde{B}v|^2 + ic[(\tilde{A}v, \tilde{B}v) - (\tilde{B}v, \tilde{A}v)] \\ &= \{(\Delta A)^2 + c^2(\Delta B)^2 + c\langle i[A, B] \rangle\}|v|^2. \end{aligned} \qquad (6.77)$$

For this inequality to hold the quadratic polynomial in c on the rhs cannot have two distinct real roots. Therefore

$$(\Delta A)^2 (\Delta B)^2 - \frac{1}{4} \langle i[A, B] \rangle^2 \geq 0. \tag{6.78}$$

In particular in $V = S(\mathbb{R})$ with $A = x \cdot$ and $B = id/dx$ we have

$$i[A, B] = \mathbb{1} \tag{6.79}$$

and the uncertainty relation of Fourier transforms follows from equations (4.31) and (4.29).

Exercises

1. Prove that the scalar product is antilinear in the first variable and non-degenerate.
2. What is the ℓ^2 norm of the sequence $(1, \frac{1}{2}, \frac{1}{3}, \frac{1}{4}, ...)$?
3. In $V = \mathbb{C}^N$ with the standard scalar product prove that
    a) a matrix U is a unitary operator if and only if U is unitary (6.56).
    b) a matrix A is a Hermitian operator if and only if A is Hermitian (6.63).
4. Is $\begin{pmatrix} 0 & i \\ -i & 0 \end{pmatrix}$ unitary? Is it Hermitian? What is its point spectrum?
5. Consider the mapping
$$\ell^2 \to \ell^2$$
$$(v^1, v^2, v^3, ...) \to (0, v^1, v^2, ...)$$
    Is it an operator? Is it unitary?
6. Show that $i\partial/\partial x^k$ is Hermitian on $C_0^\infty(\mathbb{R}^N)$ and on $S(\mathbb{R}^N)$.
7. For what functions $h(x)$ is the multiplication operator $h \cdot$ on $C_0^\infty(\mathbb{R}^N)$ Hermitian?
8. For what kernels is the integral operator (6.22) Hermitian on $C_0^\infty(\mathbb{R}^N)$?
9. Which of the following operators is Hermitian on $S(\mathbb{R}^N)$: Laplace operator, heat operator, Schrödinger operator, wave operator, Klein-Gordon operator?
10. Prove that if A and B are Hermitian then so is $i[A, B]$.
11. Show that the square root in (6.72) is well defined.
12. Compute the commutator (6.79) and derive the uncertainty relation of Fourier transforms.

6.4 Orthonormal bases

A vector space V is called N dimensional if there are N vectors $b_1, b_2, ..., b_N \in V$ such that every vector $v \in V$ can be expressed in a unique way as a linear combination of these vectors,

$$v = \sum_{j=1}^{N} a_j b_j, \quad a_j \in \mathbb{C}. \tag{6.80}$$

The vectors b_j are then called a basis. For instance the vectors

$$e_1 := \begin{pmatrix} 1 \\ 0 \\ \vdots \\ 0 \end{pmatrix}, \quad e_2 := \begin{pmatrix} 0 \\ 1 \\ 0 \\ \vdots \\ 0 \end{pmatrix}, \quad ..., \quad e_N = \begin{pmatrix} 0 \\ \vdots \\ 0 \\ 1 \end{pmatrix} \tag{6.81}$$

are a basis of $\mathbb{C}^N$, the so-called canonical basis.

There are several possible generalizations of the notion of a basis to infinite dimensional spaces.

The so-called algebraic basis of a vector space is a set b_j of vectors, where the index j runs over a not necessarily countable set J, such that every vector $v \in V$ can be written in a unique fashion as a finite linear combination of basis vectors

$$v = \sum_{k=1}^{M} a_k b_{j_k}, \quad j_k \in J. \tag{6.82}$$

For instance the monomials $1, x, x^2...$ form an algebraic basis of the space of polynomials. In general one can show, using the axiom of choice, that every vector space admits an algebraic basis. However, this basis is not very useful from a practical point of view.

We shall be interested in the so-called Hilbert or Schauder basis.

Definition: Let V be a vector space with scalar product. A countable set of vectors b_j, $j \in \mathbb{N}$ (or $j \in \mathbb{Z}$) is called (Hilbert) basis if every vector $v \in V$ can be expressed in a unique way as a linear combination of these vectors:

$$v = \sum_{j=1}^{\infty} a_j b_j, \quad a_j \in \mathbb{C} \tag{6.83}$$

where the convergence of the series is meant with respect to the norm induced by the scalar product.

A basis of ℓ^2 for instance is given by the sequences

$$e_1 := (1, 0, 0, ...), \tag{6.84}$$

$$e_2 := (0, 1, 0, ...) \tag{6.85}$$

and so forth. For the particular scalar product (6.47) convergence with respect to the norm means convergence in the mean and the harmonics $e_j(t)$ form a basis of the differentiable, periodic functions.

An extremely valuable property of a basis is orthonormality:

Definition: A Hilbert basis e_j is said to be orthonormal if its vectors are all orthogonal

$$(e_j, e_k) = 0, \qquad j \neq k, \tag{6.86}$$

and normalized to one

$$(e_j, e_j) = 1. \tag{6.87}$$

For example the bases (6.81), (6.84-5) and the harmonics are orthonormal. The following theorem reflects the usefulness of an orthonormal basis.

Generalized Fourier theorem: Let e_j, $j \in \mathbb{N}$, be an orthonormal basis of a Pre-Hilbert space V. Then for every vector $v \in V$ we have the generalized Fourier series

$$v = \sum_{j=1}^{\infty} (e_j, v) e_j \tag{6.88}$$

and the generalized Parseval equation

$$(v, w) = \sum_{j=1}^{\infty} \overline{(e_j, v)} (e_j, w), \tag{6.89}$$

in particular

$$|v|^2 = \sum_{j=1}^{\infty} |(e_j, v)|^2 \tag{6.90}$$

where all limits are understood in the sense of convergence with respect to the norm coming from the scalar product.

The proof uses a lemma:

Continuity of the scalar product: Let v_n and $w_n, n \in \mathbb{N}$ be two sequences of vectors that converge to two vectors v and w in a Pre-Hilbert space v,

$$|v_n - v| \underset{n \to \infty}{\longrightarrow} 0, \tag{6.91}$$

$$|w_n - w| \underset{n \to \infty}{\longrightarrow} 0. \tag{6.92}$$

Then the sequence of scalar products converges

$$\lim_{n \to \infty} (v_n, w_n) = (v, w). \tag{6.93}$$

Indeed using Cauchy-Schwarz and triangle inequalities we have:

$$\begin{aligned}
|(v_n, w_n) - (v, w)| &\leq |(v_n, w_n) - (v_n, w)| + |(v_n, w) - (v, w)| \\
&= |(v_n, w_n - w)| + |(v_n - v, w)| \leq |v_n||w_n - w| + |v_n - v||w| \\
&\leq |v||w_n - w| + |v_n - v||w_n - w| + |v_n - v||w| \underset{n \to \infty}{\longrightarrow} 0.
\end{aligned} \tag{6.94}$$

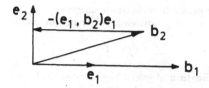

Fig. 6.2   The orthonormalization procedure by Gram and Schmidt

To prove the generalized Fourier series, we write v as a linear combination of the basis vectors

$$v = \sum_{j=1}^{\infty} a_j e_j \qquad (6.95)$$

and calculate the coefficients a_j:

$$(e_k, v) = (e_k, \lim_{n \to \infty} \sum_{j=1}^{n} a_j e_j) = \lim_{n \to \infty} (e_k, \sum_{j=1}^{n} a_j e_j)$$

$$= \lim_{n \to \infty} \sum_{j=1}^{n} a_j (e_k, e_j) = a_k \qquad (6.96)$$

because the e_j are orthonormal. The generalized Parseval equation is derived similarly. This equation is often written under the form

$$(v, w) = \sum_{j=1}^{\infty} (v, e_j)(e_j, w) \qquad (6.97)$$

or

$$\sum_{j=1}^{\infty} e_j)(e_j = \mathbb{1} \qquad (6.98)$$

and called insertion of an orthonormal basis or completeness relation. The reader should be warned, the word complete here is misleading. It has nothing to do with the same adjective used in the next section to denote the central property of a Hilbert space.

Given an arbitrary basis $b_j, j \in \mathbb{N}$, of a Pre-Hilbert space, there is a simple inductive algorithm, the Gram-Schmidt orthonormalization procedure, to obtain an orthonormal basis e_j. We set

$$e_1 := \frac{b_1}{|b_1|} \qquad (6.99)$$

and

$$e_{j+1} := \frac{b_{j+1} - \sum_{k=1}^{j} (e_k, b_{j+1}) e_k}{|b_{j+1} - \sum_{k=1}^{j} (e_k, b_{j+1}) e_k|}. \qquad (6.100)$$

It follows that the e_j are again a basis and they are orthonormal.

Of course an orthonormal basis is never unique. Applying any unitary operator U to an orthonormal basis e_j produces another orthonormal basis

$$\tilde{e}_j := Ue_j, \tag{6.101}$$

for example the multiplication of every basis vector e_j by an arbitrary "phase" z_j. A phase is a complex number with absolute value one.

We are now ready to give a practical criterium to decide if a set of vectors actually forms a basis.

Theorem: A linearly independent set of vectors b_j, $j \in \mathbb{N}$, is a basis of a Pre-Hilbert space V if and only if the zero vector is the only vector in V to be orthogonal to all b_j.

We recall that a set of vectors b_j is called linearly independent if the vanishing of a finite linear combination

$$\sum_{k=1}^{n} a_{j_k} b_{j_k} = 0 \tag{6.102}$$

implies that already all coefficients a_{j_k} vanish.

Proof: We apply the orthonormalization procedure by Gram and Schmidt to the vectors b_j and denote by e_j the resulting orthonormal vectors. The e_j are a basis of V if and only if the b_j are a basis and for a vector $v \in V$

$$(e_j, v) = 0 \quad \text{for all} \quad j \in \mathbb{N} \tag{6.103}$$

is equivalent to

$$(b_j, v) = 0 \quad \text{for all} \quad j \in \mathbb{N}. \tag{6.104}$$

Now let us assume that the e_j are a basis of V and that for a vector v all scalar products (e_j, v) vanish. Then the generalized Fourier series implies that v is already zero. Reciprocally let us assume the existence of a non-zero vector v such that all scalar products (e_j, v) vanish. Then the e_j cannot be a basis because

$$v = \sum_{k=1}^{\infty} a_k e_k \tag{6.105}$$

would imply

$$(e_j, v) = a_j = 0 \tag{6.106}$$

for all j by the continuity of the scalar product.

As an application we shall construct a basis of $\mathcal{S}(\mathbb{R})$. Let $\rho(x)$ be a differentiable, positive function with exponential fall off at plus and minus infinity. That means there are two positive constants a and C such that

$$\rho(x) < Ce^{-a|x|}. \tag{6.107}$$

Such a $\rho(x)$ is called weight function. We want to prove that the functions of fast decrease

$$b_j(x) := x^j \rho(x), \quad j = 0, 1, 2, 3, \ldots \tag{6.108}$$

form a basis of $\mathcal{S}(\mathbb{R})$. Linear independence of the b_j is easy to verify. Let f be a function of fast decrease such that all scalar products (b_j, f) vanish. We have to show that then f is already zero. To this end consider the Laplace transform

$$F(z) := \mathcal{L}(\rho f)(z) := \left(e^{-\lambda x} \rho(x) f(x)\right)\hat{\ }(k). \tag{6.109}$$

By the theorem of section 4.6, $F(z)$ is holomorphic in the strip of the complex plane defined by

$$z = \lambda - ik \in (-a, a) \times \mathbb{R}. \tag{6.110}$$

Using equations (1.48) and (4.32) we obtain

$$\left(\frac{d}{dz}\right)^j F(0) = \left(\frac{-1}{i}\frac{\partial}{\partial k}\right)^j F(0) = \widehat{b_j f}(0) = \frac{(b_j, f)}{\sqrt{2\pi}} = 0. \tag{6.111}$$

Therefore, $F(z)$ being analytic, it must be zero identically, in particular we have

$$\widehat{\rho f} \equiv 0 \tag{6.112}$$

and consequently

$$\rho f = 0. \tag{6.113}$$

By applying the orthonormalization procedure to the $b_j(x)$, we obtain for any weight function $\rho(x)$ an orthonormal basis of functions $e_j(x)$ and thereby a generalized Fourier series for functions of fast decrease. Since the functions $b_j(x)$ all contain a factor $\rho(x)$, the $e_j(x)$ as well contain this factor and $e_j(x)/\rho(x)$ are again polynomials. These polynomials now are orthonormal with respect to the modified scalar product

$$(f, g) := \int_{-\infty}^{+\infty} \bar{f}(x) g(x) \rho^2(x) dx. \tag{6.114}$$

With the choice

$$\rho(x) = e^{-x^2/2} \tag{6.115}$$

the functions $e_j(x)$ are the so-called Hermite functions that appear as eigenvectors of the differential operator with non-constant coefficients $A = -d^2/dx^2 + x^2$ which describes the harmonic oscillator in quantum mechanics. The corresponding polynomials $e_j(x)/\rho(x)$ are the Hermite polynomials.

The above argument can be generalized to prove that $b_j(x) = x^j \rho(x)$ is a basis of $\mathcal{S}([0, \infty])$, the functions of fast decrease on the positive half axis. Also they are a basis of $\mathcal{C}^\infty([-1, +1])$. In the first case the weight function is a differentiable function on $[0, \infty)$ with exponential or faster fall-off at plus infinity, in the second case ρ is any differentiable function on the closed interval $[-1, +1]$. The generalization proceeds by simply extending all functions under consideration to functions of fast decrease on the entire x-axis. In this case we arrive at the following useful systems of polynomials:

- the Laguerre polynomials for $\rho(x) = e^{-x/2}$ on the positive half axis.
- the Legendre polynomials for $\rho(x) = 1$ on $[-1, +1]$,
- the Jacobi polynomials with even index α and β for $\rho(x) = (1-x)^{\alpha/2}(1+x)^{\beta/2}$ on $[-1, +1]$.

Exercises

1. Prove the generalized Parseval equation.
2. Prove the orthonormalization procedure by Gram and Schmidt:
   a) Show that for any natural number N the vectors $e_1, e_2, ..., e_N$ span the same subspace as $b_1, b_2, ..., b_N$.
   b) Show that the e_j are orthogonal.
3. In $V = \mathbb{C}^3$ apply the orthonormalization procedure to
$$\begin{pmatrix} 1 \\ 0 \\ i \end{pmatrix}, \begin{pmatrix} 0 \\ 1 \\ -1 \end{pmatrix} \text{ and } \begin{pmatrix} i \\ 0 \\ 0 \end{pmatrix}.$$
4. Correct the proof of the last theorem.
5. Let $\rho(x)$ be any weight function. Show that the functions $b_j(x) = x^j \rho(x), j = 0, 1, 2, 3, \ldots$ are linearly independent.
6. Compute the first three Hermite functions and their eigenvalues with respect to $-d^2/dx^2 + x^2$.
7. Compute the first three Laguerre and Legendre polynomials.

6.5 Hilbert spaces

So then what is wrong with a Pre-Hilbert space? In general it is not complete in the same sense as the rational numbers have gaps that are filled in with real numbers. Consider for instance the Pre-Hilbert space of differentiable periodic functions C_0^∞(circle) with the harmonics $e_j(t)$ as orthonormal basis. We know from section 3.4 that their linear combination with coefficients

$$c_0 = 0, \qquad\qquad (6.116)$$

$$c_j = \frac{1}{2\pi i j}, \quad j \neq 0 \qquad\qquad (6.117)$$

converges in the mean to the harmonic saw which does not belong to C_0^∞(circle). In other words, the partial sums of the Fourier series of the harmonic saw are a Cauchy sequence in the Pre-Hilbert space C_0^∞(circle) with respect to convergence in the mean. However, this Cauchy sequence does not converge to an element in this space . We are led to look for a bigger space, the completion of the Pre-Hilbert space. Let us recall that we had introduced the distributions by completing the differentiable functions with respect to weak convergence. But $\mathcal{D}(\text{circle})$ is not the bigger space we are looking for now because the scalar product (6.47) cannot be extended to all periodic distributions. On the other hand every Cauchy sequence $f_n(t)$ with respect to convergence in the mean in $C_0^\infty(\text{circle})$ defines a unique periodic distribution $D(t)$. Indeed for any test function $T(t)$ we have:

$$\left| \int_0^T f_n(t)T(t)dt - \int_0^T f_m(t)T(t)dt \right|$$
$$= |(\bar{T}, f_n - f_m)| \leq |T||f_n - f_m|. \tag{6.118}$$

Reciprocally two equivalent Cauchy sequences yield the same distribution. Finally by Parseval's theorem (6.90) the completion of $C_0^\infty(\text{circle})$ is the space of those periodic distributions whose Fourier coefficients c_j are square summable

$$\sum_{j=-\infty}^{+\infty} |c_j|^2 < \infty. \tag{6.119}$$

We denote this space by $\mathcal{L}^2(\text{circle})$. It is almost the space of periodic square integrable functions. Almost, because any function that vanishes everywhere except in one point is square integrable and its norm is zero, contradicting the positivity axiom (6.43). Therefore any such function is to be identified with zero which in the sense of distributions is already done.

Definition: A (separable) Hilbert space is a complex vector space equipped with a scalar product which admits a (countable) basis and is complete with respect to the norm of the scalar product.

Of course any finite dimensional Pre-Hilbert space is complete. Also ℓ^2 is already complete and therefore a Hilbert space. By construction $\mathcal{L}^2(\text{circle})$ is complete and the Fourier series defines an isomorphism between these two Hilbert spaces.

Similarly we denote by $\mathcal{L}^2(\Omega)$ the completion of $C_0^\infty(\Omega)$. As with rational and real numbers, $C_0^\infty(\Omega)$ is dense in $\mathcal{L}^2(\Omega)$ with respect to the norm of the scalar product. Every element of $\mathcal{L}^2(\mathbb{R}^N)$ is a tempered distribution and can be represented by a square integrable function. Contrary to tenacious rumours in physicist circles square integrable functions do not necessarily fall off at infinity. For example the function

$$f(x) = x^2 e^{-x^8 \sin^2 x} \tag{6.120}$$

although unbounded is an honest element of $\mathcal{L}^2(\mathbb{R})$. On the other hand if all partial derivatives of order less than or equal to $N/2 + 1$ of a distribution in $\mathcal{L}^2(\mathbb{R}^N)$ are again in $\mathcal{L}^2(\mathbb{R}^N)$, then this distribution is a continuous function which goes to zero with increasing radius $r = |x|$.

At the end of the last section we presented an algorithm to construct orthonormal bases by orthonormalizing the functions $x^j \rho(x)$, $j = 0, 1, 2, 3, \ldots$. This procedure remains valid in the Hilbert space $\mathcal{L}^2([-1, 1])$ if the weight function $\rho(x)$ is square integrable. This algorithm then produces further systems of polynomials:

- the Tschebysheff polynomials of first kind for $\rho(x) = (1 - x^2)^{-1/4}$,
- the Tschebysheff polynomials of second kind for $\rho(x) = (1 - x^2)^{1/4}$
- the Gegenbauer polynomials for $\rho(x) = (1 - x^2)^{(2\lambda-1)/4}, \lambda > -1/2$.

Exercises

1. Show that ℓ^2 is complete.
2. Is $\mathcal{S}(\mathbb{R})$ a Hilbert space?
3. Show that $f(x) = x^2 e^{-x^8 \sin^2 x}$ is square integrable on the real line.
4. Show that $\rho(x) = (1 - x^2)^{(2\lambda-1)/4}$ is in $\mathcal{L}^2([-1, +1])$ for $\lambda > -1/2$.
5. Compute three Tschebysheff polynomials of first and of second kind.

6.6 Self-adjoint operators

Operators have much more difficult lives in a Hilbert space than in a Pre-Hilbert space. For instance in $\mathcal{C}_0^\infty(\text{circle})$ the derivative is well defined everywhere, not so in
$\mathcal{L}^2(\text{circle})$: The derivative of the harmonic saw, although well defined as distribution, is not an element of $\mathcal{L}^2(\text{circle})$. In general we have to be content with a weaker definition of operators on a Hilbert space.

Definition: An operator A on a Hilbert space V consists of two informations, first the domain of definition $\mathcal{D}(A)$, a subspace of V, second a linear map from this subspace into V.

We have already seen how decisive the domain of definition was for various properties of an operator. We also need a more general notion of eigenvalue.

Definition: Let A be an operator on a Hilbert space V. We call a complex number λ generalized eigenvalue of A if for any positive ϵ there is an approximate eigenvector $v \in \mathcal{D}(A)$, i. e. a vector with norm $|v| \geq 1$ such that

$$|Av - \lambda v| < \epsilon. \tag{6.121}$$

The set of all generalized eigenvalues is called the spectrum of A.

Of course every eigenvalue is also a generalized eigenvalue and hence the spectrum contains the point spectrum.

Examples: Consider the Hilbert space $\mathcal{L}^2(\mathbb{R})$, the domain of definition

$$\mathcal{D}(A) := \{f(x) \in \mathcal{L}^2(\mathbb{R}) | xf(x) \in \mathcal{L}^2(\mathbb{R})\} \tag{6.122}$$

and the multiplication operator $A = x$. Its point spectrum is empty, its spectrum is the entire real axis. Indeed to every real λ the Gaussian strongly peaked at $x = \lambda$

$$f_\lambda(x) = \frac{1}{2\epsilon\sqrt{\pi}} e^{-(x-\lambda)^2/(4\epsilon^2)} \tag{6.123}$$

is an approximate eigenvector for any positive small ϵ.

Next consider the operator $A = \mathrm{i}d/dx$ with domain of definition

$$\mathcal{D}(A) := \{f \in \mathcal{L}^2(\mathbb{R}) | f' \in \mathcal{L}^2(\mathbb{R})\} \tag{6.124}$$

as subspace of $\mathcal{L}^2(\mathbb{R})$. Again its point spectrum is empty, its spectrum is the entire real axis. Approximate eigenvectors are the well-known wave packets, Fourier transforms of (6.123),

$$f_\lambda(x) = \frac{1}{\sqrt{2\pi}} e^{-\epsilon^2 x^2} e^{-\mathrm{i}\lambda x}. \tag{6.125}$$

We now define the adjoint $A^\dagger$ of an operator A on a Hilbert space V. We must require that the domain of definition $\mathcal{D}(A)$ is dense in V. Then the domain of definition $\mathcal{D}(A^\dagger)$ is the set of all vectors $v \in V$ such that there is a vector $w \in V$ satisfying

$$(v, Au) = (w, u) \tag{6.126}$$

for all u in $\mathcal{D}(A)$. If such a vector w exists for given v, then w is unique and we set

$$A^\dagger v := w. \tag{6.127}$$

For example, in finite dimensional Hilbert spaces $V = \mathbb{C}^N$ the adjoint is given by

$$A^\dagger = \bar{A}^T. \tag{6.128}$$

If U is a unitary operator on any Hilbert space V, then by definition

$$\mathcal{D}(U) = \mathcal{D}(U^{-1}) = V \tag{6.129}$$

and we have

$$U^\dagger = U^{-1} \tag{6.130}$$

as for example the Fourier transform on $\mathcal{L}^2(\mathbb{R})$.

Definition: An operator A on a Hilbert space V is called self-adjoint if its domain of definition $\mathcal{D}(A)$ is dense in V and if A coincides with its adjoint:

$$\mathcal{D}(A^\dagger) = \mathcal{D}(A), \tag{6.131}$$

$$A^\dagger = A. \tag{6.132}$$

In particular every self-adjoint operator is Hermitian in its domain of definition. On the other hand here is an operator that is Hermitian in its domain of definition, but not self-adjoint. Our example is the radial momentum operator which lives on $\mathcal{L}^2([0, \infty))$:

$$\mathcal{D}(A) = \{f \in \mathcal{L}^2([0, \infty)) | f' \in \mathcal{L}^2([0, \infty))\} \tag{6.133}$$

$$Af = \mathrm{i}f' \tag{6.134}$$

whose adjoint is given by

$$\mathcal{D}(A^\dagger) = \{f \in \mathcal{L}^2([0,\infty))|f' \in \mathcal{L}^2([0,\infty)), f(0) = 0\} \tag{6.135}$$

$$A^\dagger f = \mathrm{i}f' \tag{6.136}$$

Important self-adjoint operators are x and $\mathrm{i}d/dx$ with domains of definition (6.122) and (6.124), which are dense because already the test functions are dense.

We already know that every eigenvalue of a self-adjoint operator is real. This remains true for generalized eigenvalues. Furthermore eigenvectors with different eigenvalues are orthogonal. Therefore the point spectrum of a self-adjoint operator is countable. Its complement in the spectrum is called continuous spectrum.

A fundamental theorem of linear algebra, the diagonalization of Hermitian matrices, states that in a finite dimensional Hilbert space every self-adjoint operator admits an orthonormal basis of eigenvectors. The generalization of this theorem to infinite dimensional Hilbert spaces is a fundamental theorem of functional analysis, the spectral decomposition of a self-adjoint operator. Its difficult part lies in the continuous spectrum where a continuous family of approximate eigenvectors has to be used instead of a basis. If the continuous spectrum is empty, then an orthonormal set of eigenvectors e_λ with eigenvalues λ forms a basis and the spectral decomposition is the generalized Fourier series (6.88)

$$v = \sum_{\substack{point \\ spectrum}} (e_\lambda, v)e_\lambda. \tag{6.137}$$

An example here is the self-adjoint operator $\mathrm{i}d/dt$ on $\mathcal{L}^2(\text{circle})$ with eigenvalues $j2\pi/T, \quad j \in \mathbb{Z}$. Its spectral decomposition is the Fourier series.

On the other hand, let us consider a self-adjoint operator with purely continuous spectrum, the multiplication operator $x\cdot$. Its approximate eigenvectors $f_\lambda(x)$ from equation (6.123) form a continuous family indexed by $\lambda \in \mathbb{R}$. It is tempting to try and get rid of ϵ by sending it to zero. This limit fails to converge in the mean. However, according to equation (4.49) it does converge weakly:

$$f_\lambda(x) = \frac{1}{2\epsilon\sqrt{\pi}}e^{-(x-\lambda)^2/(4\epsilon^2)} \xrightarrow[\epsilon \to 0]{} \delta(x - \lambda) =: e_\lambda(x). \tag{6.138}$$

This limit, which is not in $\mathcal{L}^2(\mathbb{R})$, is called eigendistribution with eigenvalue λ. In this case the spectral decomposition reads

$$v(x) = \int_{\substack{continuous \\ spectrum}} \bar{e}_\lambda(v)e_\lambda d\lambda \tag{6.139}$$

where $v(x)$ is in $\mathcal{L}^2(\mathbb{R})$ and

$$\bar{e}_\lambda(v) := \int_{-\infty}^{+\infty} \bar{e}_\lambda(x)v(x)dx = v(\lambda) \tag{6.140}$$

is the evaluation of the eigendistribution on the function $v(x)$. Although $v(x)$ is not a test function, this expression has a meaning because the test functions are dense in $\mathcal{L}^2(\mathbb{R})$. Finally the spectral decomposition here reduces to

$$v(x) = \int_{-\infty}^{+\infty} v(\lambda)\delta(x-\lambda)d\lambda = (\delta * v)(x) \qquad (6.141)$$

which is nothing but equation (2.63) extended to $\mathcal{L}^2(\mathbb{R})$.

Now let us consider the self-adjoint operator id/dx. Again the spectrum is purely continuous with eigendistributions

$$f_\lambda(x) = \frac{1}{\sqrt{2\pi}}e^{-\epsilon^2 x^2}e^{-i\lambda x} \xrightarrow[\epsilon \to 0]{} \frac{1}{\sqrt{2\pi}}e^{-i\lambda x} =: e_\lambda(x). \qquad (6.142)$$

i. e. the plane waves. Therefore

$$\bar{e}_\lambda(v) = \int_{-\infty}^{+\infty} \frac{1}{\sqrt{2\pi}}e^{i\lambda x}v(x)dx = \hat{v}(\lambda) \qquad (6.143)$$

and the spectral decomposition is Fourier's inversion formula in $\mathcal{L}^2(\mathbb{R})$

$$v(x) = \int_{-\infty}^{+\infty} \hat{v}(\lambda)\frac{1}{\sqrt{2\pi}}e^{-i\lambda x}dx. \qquad (6.144)$$

A general self-adjoint operator has both point and continuous spectrum corresponding to bound states and scattering states in quantum mechanics. In this case the spectral decomposition contains a sum and an integral. The general theorem, in particular the normalization of the eigendistributions, is beyond the scope of this text. The natural setting of the spectral decomposition is the rigged Hilbert space or Gelfand triple, which was introduced around 1960 by Gelfand and collaborators. A rigged Hilbert space consists of three nested spaces

$$\mathcal{P} \subset \mathcal{H} \subset \mathcal{G}. \qquad (6.145)$$

The first is a Pre-Hilbert space $\mathcal{P}$ that in addition to its convergence with respect to the scalar product has a second weaker notion of convergence, e.g. weak convergence. The second space $\mathcal{H}$, the completion of $\mathcal{P}$ with respect to the scalar product, is a Hilbert space. The third is the completion of $\mathcal{P}$ with respect to the weaker convergence. We have already encountered two rigged Hilbert spaces

$$\mathcal{C}_0^\infty(\Omega) \subset \mathcal{L}^2(\Omega) \subset \mathcal{D}(\Omega) \qquad (6.146)$$

138

and

$$\mathcal{S}(\mathbb{R}^N) \subset \mathcal{L}^2(\mathbb{R}^N) \subset \mathcal{T}(\mathbb{R}^N). \qquad (6.147)$$

The eigendistributions are elements of the biggest space, which does not carry a scalar product.

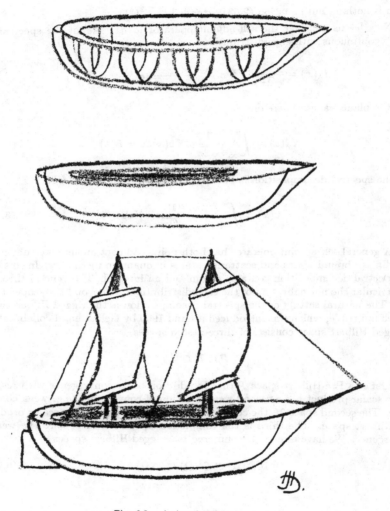

Fig. 6.3   A rigged Hilbert space

Exercises

1. Exhibit a square integrable function $f(x)$ such that $xf(x)$ is not square integrable.
2. Exhibit a square integrable function $f(x)$ such that $f'(x)$ is a well defined function but not square integrable.
3. Show that the functions (6.123) and (6.125) are approximate eigenvectors.
4. Show that the vector w in equation (6.126) is unique if it exists.
5. In ℓ^2 consider the annihilation and creation operators a and $a^\dagger$ with domain of definition

$$\mathcal{D}(a) = \mathcal{D}(a^\dagger) = \{v = (v^0, v^1, v^2, ...), \sum_{j=0}^{\infty} j|v^j|^2 < \infty\}$$

defined by

$$av := (v^1, \sqrt{2}v^2, \sqrt{3}v^3, ...)$$
$$a^\dagger v := (0, v^0, \sqrt{2}v^1, \sqrt{3}v^2, ...).$$

Show that $a^\dagger$ is the adjoint of a. Show that the point spectrum of a is the entire complex plane, the point spectrum of $a^\dagger$ is empty. Find a domain of definition for the particle number operator

$$N = a^\dagger a$$

that makes N self-adjoint. On this domain prove the commutation relation

$$[a^\dagger, a] = -1 .$$

Find an orthonormal basis of eigenvectors of N.
6. Let A be an operator with dense domain of definition in a Hilbert space. Is $A^{\dagger\dagger} = A$?
7. Compute the adjoint of the radial momentum operator (6.133-4).
8. For what differentiable functions $h(x)$ is the multiplication operator $h\cdot$ self-adjoint on $\mathcal{L}^2(\mathbb{R}^N)$?

7

Systems of special functions

The tremendous power of the Fourier series relies on two properties of the harmonics. They are an orthonormal basis and they are eigenvectors of the self-adjoint differential operator id/dt. According to the spectral decomposition the Fourier series only scratches the surface of a gold mine. The purpose of the present chapter is to illustrate this wealth. We shall indicate the spectral decomposition of a few linear differential operators that appear in physics. These differential operators all have non-constant coefficients and could not be treated with the techniques of chapter 5. In most of our examples the operator has a pure point spectrum so that its eigenvectors are functions automatically. It should be noted that we introduce special functions only from the view point of functional analysis. Our discussion does not do justice to the group-theoretical view point which is of equally fundamental importance in physics. For the group-theoretical approach we recommend the book by Talman (1968).

7.1 Hermite functions and polynomials

Our first example is the differential operator of the harmonic oscillator in quantum mechanics:

$$A := -\frac{d^2}{dx^2} + x^2. \tag{7.1}$$

With the domain of definition

$$\mathcal{D}(A) := \{f \in \mathcal{L}^2(\mathbb{R}) \mid Af \in \mathcal{L}^2(\mathbb{R})\} \tag{7.2}$$

this operator is self-adjoint. Its pure point-spectrum is $\lambda = 1, 3, 5....$ The corresponding orthogonal eigenfunctions are the Hermite functions $h_j(x)$ with $j = (\lambda - 1)/2$:

$$h_0(x) := e^{-x^2/2}, \tag{7.3}$$

$$h_1(x) := 2xe^{-x^2/2}, \tag{7.4}$$

$$h_2(x) := (4x^2 - 2)e^{-x^2/2}. \tag{7.5}$$

In general the Hermite functions are given by

$$h_j(x) = (x - \frac{d}{dx})^j e^{-x^2/2}, \quad j = 0, 1, 2, ... \tag{7.6}$$

They are orthogonal but not normalized to one:

$$|h_j|^2 = \int_{-\infty}^{+\infty} h_j^2(x)dx = \sqrt{\pi}j!2^j. \tag{7.7}$$

The reader should be warned that the normalization of orthogonal polynomials is not uniform in the literature.

The Gram-Schmidt procedure determines the j plus first orthogonal basis vector as a linear combination (with constant coefficients) of all previous ones, equation (6.100). Here the following recurrence relation with non-constant coefficients is of practical use:

$$h_{j+1} = 2xh_j - 2jh_{j-1}. \tag{7.8}$$

As already mentioned, we obtain the Hermite polynomials by dividing the corresponding functions by the weight function $\rho(x) = e^{-x^2/2}$. Therefore we are led to define a new Hilbert space

$$\mathcal{L}^2_{\rho^2}(\mathbb{R}) := \{f : \mathbb{R} \to \mathbb{C}, \int |f|^2\rho^2 < \infty\} \tag{7.9}$$

with scalar product

$$(f,g) := \int \bar{f}g\rho^2. \tag{7.10}$$

With respect to this scalar product the operator

$$A := -\frac{d^2}{dx^2} + 2x\frac{d}{dx} + 1 \tag{7.11}$$

is self-adjoint and admits an orthogonal basis of eigenvectors, the Hermite polynomials

$$H_j(x) := h_j(x)e^{x^2/2} \tag{7.12}$$

again with eigenvalues $\lambda = 2j + 1$. Also the Hermite polynomials have the same norm (with respect to the scalar product (7.10)) and the same recurrence relation (7.8) as the Hermite functions.

Exercises

1. Show that the functions $\left(x - \frac{d}{dx}\right)^j x^{-x^2/2}$ with $j = 0, 1, 2, \dots$ are eigenfunctions of $-d^2/dx^2 + x^2$ with eigenvalues $2j + 1$. Show that they form a basis of $\mathcal{L}^2(\mathbb{R})$.
2. Prove the recurrence relation (7.8).
3. Show that the Hermite functions are eigenvectors of the Fourier transform and compute their eigenvalues.
4. Prove that the operator (7.11) is Hermitian on the Pre-Hilbert space of polynomials with scalar product (7.10). With respect to what domain of definition is this operator self-adjoint?

5. Show that the Hermite polynomials can be calculated by means of the formula

$$H_j(x) = e^{x^2} \left(-\frac{d}{dx} \right)^j e^{-x^2}.$$

6. Show that the Hermite polynomials are eigenvectors of the unitary operator "parity"

$$(Pf)(x) = f(-x)$$

and compute their eigenvalues.

7.2 Other orthogonal polynomials

Following the same pattern we indicate differential operators belonging to other polynomials we have encountered. In each case the Hilbert space is $\mathcal{L}^2_{\rho^2}(\Omega)$ and the spectrum a pure point spectrum.

The Laguerre polynomials L_j:

$$\Omega = [0, \infty) \tag{7.13}$$

$$\rho(x) = e^{-x/2} \tag{7.14}$$

$$A = -x\frac{d^2}{dx^2} + (x-1)\frac{d}{dx} \tag{7.15}$$

$$L_j(x) = e^x \left(\frac{d}{dx} \right)^j (x^j e^{-x}), \quad j = 0,1,2,\ldots \tag{7.16}$$

$$\lambda = j \tag{7.17}$$

$$|L_j|^2 := \int_0^\infty L_j^2(x) e^{-x} dx = (j!)^2 \tag{7.18}$$

$$L_{j+1}(x) = (2j+1-x)L_j(x) - j^2 L_{j-1}(x) \tag{7.19}$$

$$L_0(x) = 1 \tag{7.20}$$

$$L_1(x) = 1 - x \tag{7.21}$$

$$L_2(x) = 2 - 4x + x^2 \tag{7.22}$$

The Legendre polynomials P_j:

$$\Omega = [-1, 1] \tag{7.23}$$

$$\rho(x) \equiv 1 \tag{7.24}$$

$$A = -(1-x^2)\frac{d^2}{dx^2} + 2x\frac{d}{dx} \tag{7.25}$$

$$P_j(x) = \frac{1}{j!} \left(\frac{1}{2}\frac{d}{dx} \right)^j (x^2-1)^j, \quad j = 0,1,2,\ldots \tag{7.26}$$

$$\lambda = j(j+1) \tag{7.27}$$

$$|P_j|^2 = \int_{-1}^{1} P_j^2(x)dx = \frac{2}{2j+1} \tag{7.28}$$

$$P_{j+1}(x) = \frac{2j+1}{j+1}xP_j(x) - \frac{j}{j+1}P_{j-1}(x) \tag{7.29}$$

$$P_0(x) = 1 \tag{7.30}$$

$$P_1(x) = x \tag{7.31}$$

$$P_2(x) = \frac{1}{2}(3x^2 - 1) \tag{7.32}$$

The Tschebysheff polynomials of first kind $T_j(x)$:

$$\Omega = [-1, 1] \tag{7.33}$$

$$\rho(x) = (1 - x^2)^{-1/4} \tag{7.34}$$

$$A = -(1 - x^2)\frac{d^2}{dx^2} + x\frac{d}{dx} \tag{7.35}$$

$$T_j(x) = (1 - x^2)^{1/2}\frac{1}{1 \cdot 3 \cdot 5 \cdot \ldots (2j-1)}\left(-\frac{d}{dx}\right)^j (1 - x^2)^{j-1/2}, \quad j = 0, 1, 2, \ldots \tag{7.36}$$

$$\lambda = j^2 \tag{7.37}$$

$$|T_j|^2 = \int_{-1}^{1} T_j^2(x)\frac{dx}{\sqrt{1-x^2}} = \begin{cases} \pi, & j = 0 \\ \pi/2, & j = 1, 2, 3, \ldots \end{cases} \tag{7.38}$$

$$T_{j+1}(x) = 2xT_j(x) - T_{j-1}(x) \tag{7.39}$$

$$T_0(x) = 1 \tag{7.40}$$

$$T_1(x) = x \tag{7.41}$$

$$T_2(x) = 2x^2 - 1 \tag{7.42}$$

The Tschebysheff polynomials of second kind $U_j(x)$:

$$\Omega = [-1, 1] \tag{7.43}$$

$$\rho(x) = (1 - x^2)^{1/4} \tag{7.44}$$

$$A = -(1 - x^2)\frac{d^2}{dx^2} + 3x\frac{d}{dx} \tag{7.45}$$

$$U_j(x) = (1 - x^2)^{-1/2}\frac{j+1}{1 \cdot 3 \cdot 5 \cdot \ldots (2j+1)}\left(-\frac{d}{dx}\right)^j (1 - x^2)^{j+1/2}, \quad j = 0, 1, 2, \ldots \tag{7.46}$$

144

$$\lambda = j(j+2) \tag{7.47}$$

$$|T_j|^2 = \int_{-1}^{1} U_j^2(x)\sqrt{1-x^2}dx = \frac{\pi}{2} \tag{7.48}$$

$$U_{j+1}(x) = 2xU_j(x) - U_{j-1}(x) \tag{7.49}$$

$$U_0(x) = 1 \tag{7.50}$$

$$U_1(x) = 2x \tag{7.51}$$

$$U_2(x) = 4x^2 - 1 \tag{7.52}$$

Exercises

1. What differential operator belongs to the Laguerre functions
$\ell_j(x) = L_j(x)e^{-x/2}$?
2. Show that the Tschebysheff functions of second kind $u_j(x) := U_{j-1}(x)\sqrt{1-x^2}$,
$j = 1, 2, 3, \ldots$ are eigenfunctions of the differential operator (7.35). What are
their eigenvalues?

7.3 The spherical harmonics

For each Legendre polynomial $P_j(x)$ we define the associated Legendre functions

$$P_j^m(x) := (1-x^2)^{m/2}\left(-\frac{d}{dx}\right)^m P_j(x), \quad m = 0, 1, 2, \ldots, j \tag{7.53}$$

They are eigenfunctions of the differential operator

$$A = -(1-x^2)\frac{d^2}{dx^2} + 2x\frac{d}{dx} + \frac{m^2}{1-x^2} \tag{7.54}$$

with eigenvalues

$$\lambda = j(j+1). \tag{7.55}$$

They are all orthogonal to each other,

$$\int_{-1}^{1} P_j^m(x)P_{j'}^{m'}(x)dx = \frac{2}{2j+1}\frac{j+m)!}{(j-m)!}\delta_{jj'}\delta_{mm'} \tag{7.56}$$

and satisfy the recurrence relations

$$P_j^{m+2}(x) = -2(m+1)x(1-x^2)^{-1/2}P_j^{m+1}(x) - (j-m)(j+m+1)P_j^m(x),$$
$$0 \le m \le j-2, \tag{7.57}$$

$$P_{j+1}^m(x) = \frac{2j+1}{j-m+1} x P_j^m(x) - \frac{j+m}{j-m+1} P_{j-1}^m(x), \tag{7.58}$$
$$0 \le m \le j - 1,$$

$$P_1^1(x) = -\sqrt{1-x^2}, \tag{7.59}$$

$$P_2^1(x) = -3x\sqrt{1-x^2}, \tag{7.60}$$

$$P_2^2(x) = 3x(1-x^2), \tag{7.61}$$

$$P_3^1(x) = -\frac{3}{2}(5x^2 - 1)\sqrt{1-x^2}, \tag{7.62}$$

$$P_3^2(x) = 15x(1-x^2), \tag{7.63}$$

$$P_3^3(x) = -15(1-x^2)\sqrt{1-x^2}. \tag{7.64}$$

Consider the Laplace operator on the unit sphere in $\mathbb{R}^3$. In polar coordinates (2.116-118) the sphere is parametrized by the two angles ϕ and θ and the Laplace operator on the sphere is obtained from the Laplace operator in $\mathbb{R}^3$,

$$\Delta = \frac{\partial^2}{\partial r^2} + \frac{2}{r}\frac{\partial}{\partial r} + \frac{1}{r^2 \sin^2\theta}\frac{\partial^2}{\partial \phi^2} + \frac{1}{r^2}\frac{\partial^2}{\partial \theta^2} + \frac{\cot\theta}{r^2}\frac{\partial}{\partial \theta} \tag{7.65}$$

by putting $r = 1$.

$$\Delta = \frac{1}{\sin^2\theta}\frac{\partial^2}{\partial \phi^2} + \frac{\partial^2}{\partial \theta^2} + \cot\theta\frac{\partial}{\partial \theta}. \tag{7.66}$$

After the change of variable

$$z := \cos\theta \tag{7.67}$$

with

$$\frac{\partial}{\partial \theta} = -\sqrt{1-z^2}\frac{\partial}{\partial z}, \tag{7.68}$$

$$\frac{\partial^2}{\partial \theta^2} = (1-z^2)\frac{\partial^2}{\partial z^2} - z\frac{\partial}{\partial z} \tag{7.69}$$

the Laplace operator takes the form

$$\Delta = (1-z^2)\frac{\partial^2}{dz^2} - 2z\frac{\partial}{\partial z} + \frac{1}{1-z^2}\frac{\partial}{\partial \phi^2} \tag{7.70}$$

and a separation of variables

$$f(z,\phi) = F(z)G(\phi) \tag{7.71}$$

leads us to the operator (7.54) in z. Therefore the functions

$$Y_\ell^m(\theta,\phi) = (-1)^m \left[\frac{2\ell+1}{4\pi}\frac{(\ell-m)!}{(\ell+m)!}\right]^{1/2} P_\ell^m(\cos\theta)e^{im\phi} \tag{7.72}$$

with $\ell = 0, 1, 2, \ldots$ and $m = -\ell, -\ell + 1, \ldots, \ell - 1, \ell$ are an orthonormal basis of $\mathcal{L}^2$(sphere) consisting of eigenvectors to the Laplace operator with eigenvalues

$$\lambda = -\ell(\ell + 1). \tag{7.73}$$

In formulas:

$$(Y_\ell^m, Y_{\ell'}^{m'}) = \int_0^{2\pi} \int_0^{\pi} \bar{Y}_\ell^m(\theta, \phi) Y_{\ell'}^{m'}(\theta, \phi) \sin\theta d\theta d\phi = \delta_{\ell\ell'}\delta_{mm'}, \tag{7.74}$$

$$\Delta Y_\ell^m = -\ell(\ell + 1)Y_\ell^m. \tag{7.75}$$

Note that the spectrum of the Laplace operator is degenerate: for fixed ℓ we have $2\ell + 1$ linearly independent eigenvectors with eigenvalue $-\ell(\ell + 1)$. These are simultaneously eigenvectors of the self-adjoint operator $\frac{1}{i}\frac{\partial}{\partial\phi}$ with distinct eigenvalues m,

$$\frac{1}{i}\frac{\partial}{\partial\phi}Y_\ell^m = mY_\ell^m. \tag{7.76}$$

The functions $Y_\ell^m(\theta, \phi)$ are called spherical harmonics although only $Y_0^0(\theta, \phi)$, which is constant, is a harmonic function. The spherical harmonics appear in many different branches of physics. Well known examples are the multipoles of a radiating antenna in electrodynamics or in quantum mechanics the wave function of a particle subject to a spherically symmetric potential. The next section illustrates an application to hydrodynamics.

Exercises

1. Show that the associated Legendre functions are eigenfunctions of the operator (7.54).
2. Prove equations (7.68-70).
3. Show that the function on [-1,+1]

$$G_\xi(x) := \begin{cases} \frac{1}{2j} \left(\frac{1-\xi}{1+\xi}\frac{1+x}{1-x} \right)^{j/2}, x < \xi \\[2ex] \frac{1}{2j} \left(\frac{1-x}{1+x}\frac{1+\xi}{1-\xi} \right)^{j/2}, x > \xi \end{cases}$$

is a Green function of the differential operator (7.54) for all $j = 1, 2, 3, \ldots$

7.4 Stokes' law of hydrodynamics

The asymptotic velocity v_0 of a falling sphere in an incompressible, viscous fluid is given by Stokes' law

$$F = 6\pi R\eta v_0 \tag{7.77}$$

where F ist the viscous force on the sphere, R its radius and η the viscosity. The purpose of this section is to derive Stokes' law from the equations of motion of the fluid, the Navier-Stokes equation:

$$\frac{\partial v}{\partial t} + (v \cdot \text{grad})v = -\frac{1}{\rho}\text{grad}p + \frac{\eta}{\rho}\Delta v \tag{7.78}$$

where

$$v = \begin{pmatrix} v^1 \\ v^2 \\ v^3 \end{pmatrix} \tag{7.79}$$

is the velocity vector field of the fluid in Cartesian coordinates with origin at the center of the sphere, ρ is the density of the fluid and p the pressure. Our situation allows two simplifications, the flow is stationary,

$$\frac{\partial v}{\partial t} = 0, \tag{7.80}$$

and will be assumed laminar,

$$(v \cdot \text{grad})v = 0. \tag{7.81}$$

The second assumption brings us back to linear differential equations, our task is to solve

$$\Delta v = \frac{1}{\eta}\text{grad}p \tag{7.82}$$

and the continuity equation expressing incompressibility,

$$\text{div}v = 0. \tag{7.83}$$

These are four linear equations in four unknowns v and p subject to the boundary conditions: Far away from the sphere the velocity is constant, say v_0 along the x^3-axis, and the pressure is constant, say zero. On the sphere the velocity vanishes.

Naturally we introduce polar coordinates (2.116-118). Then the axial symmetry implies that the velocity has vanishing ϕ-component v_ϕ, c.f. equation (2.123), and that the other three functions v_r, v_θ and p depend on r and θ only. Therefore the pressure can be expanded as

$$p(r, \theta) = \sum_{j=0}^{\infty} K_j(r)P_j(\cos \theta). \tag{7.84}$$

To determine the r-dependence of the coefficients K_j we use the fact that the pressure is a harmonic function:

$$\Delta p = \text{div grad}p = \eta\text{div}\Delta v = \eta\Delta\text{div}v = 0. \tag{7.85}$$

148

Then by equation (7.66)

$$\Delta(K_j(r)P_j(\cos\theta)) = (K_j'' + \frac{2}{r}K_j' - \frac{j(j+1)}{r^2}K_j)P_j = 0 \qquad (7.86)$$

implying

$$K_j(r) \propto r^j \qquad (7.87)$$

or

$$K_j(r) \propto \frac{1}{r^{j+1}}. \qquad (7.88)$$

Only the second possibility is compatible with the boundary condition for the pressure. Hence we must have

$$p = \sum_{j=0}^{\infty} C_j \frac{1}{r^{j+1}} P_j(\cos\theta), \quad C_j \in \mathbb{R}. \qquad (7.89)$$

The velocities not being harmonic cannot be written in this form. Therefore we introduce an auxiliary scalar field s such that

$$\Delta s = \frac{1}{\eta}p \qquad (7.90)$$

and define a new vector field u by

$$u := v - \text{grad}s \qquad (7.91)$$

which now is harmonic:

$$\Delta u = \Delta v - \Delta \text{grad}s = \frac{1}{\eta}\text{grad}p - \text{grad}\Delta s = 0. \qquad (7.92)$$

The auxiliary field s is not unique, any harmonic function may be added to s at will. This additional freedom suggests the ansatz

$$u = \begin{pmatrix} 0 \\ 0 \\ u^3 \end{pmatrix} \qquad (7.93)$$

which trades in the two remaining unknowns v_r and v_θ for two new unknowns s and u^3. The validity of this ansatz will become apparent later. As above u^3 being harmonic must be of the form

$$u^3 = \sum_{j=0}^{\infty}(A_j r^j + B_j \frac{1}{r^{j+1}})P_j(\cos\theta) \qquad (7.94)$$

with real coefficients A_j and B_j. The continuity equation now reads

$$\text{div}\,u = \frac{\partial}{\partial x^3}u^3 = \left(\cos\theta\frac{\partial}{\partial r} - \sin\theta\frac{1}{r}\frac{\partial}{\partial\theta}\right)u^3 = \frac{-1}{\eta}p. \tag{7.95}$$

Using properties (7.26) and (7.29) of the Legendre polynomials the divergence becomes:

$$\text{div}\,u = \sum_{j=0}^{\infty}[(j+1)A_{j+1}r^j - jB_{j-1}\frac{1}{r^{j+1}}]P_j(\cos\theta) \tag{7.96}$$

implying

$$A_j = 0, \quad j = 1,2,3,\ldots \tag{7.97}$$

$$B_j = \frac{C_{j+1}}{(j+1)\eta}. \tag{7.98}$$

Next we need a particular solution of the inhomogeneous equation (7.90). With the ansatz

$$s = \sum_{j=0}^{\infty}s_j = \sum_{j=0}^{\infty}\frac{K_j}{r^\ell}P_j(\cos\theta) \tag{7.99}$$

we obtain

$$\Delta s_j = K_j[\ell(\ell-1) - j(j+1)]\frac{1}{r^{\ell+2}}P_j = \frac{C_j}{\eta}\frac{1}{r^{j+1}}P_j \tag{7.100}$$

and consequently

$$\ell = j - 1, \tag{7.101}$$

$$K_j = \frac{C_j}{(2-4j)\eta}. \tag{7.102}$$

Finally the most general auxiliary field is

$$s = \sum_{j=0}^{\infty}\left[\frac{C_j}{(2-4j)\eta}\frac{1}{r^{j-1}} + D_j r^j + E_j\frac{1}{r^{j+1}}\right]P_j(\cos\theta). \tag{7.103}$$

So far we have copied the Fourier trick and translated all differential equations into algebraic equations for the coefficients. Now the remaining coefficients must be determined by the boundary conditions. Indeed

$$\lim_{r\to\infty}v(r,\theta) = \begin{pmatrix} 0 \\ 0 \\ v_0 \end{pmatrix} \tag{7.104}$$

yields

$$D_1 + A_0 = 0, \tag{7.105}$$

$$D_j = 0, \quad j = 2,3,4,\ldots \tag{7.106}$$

150

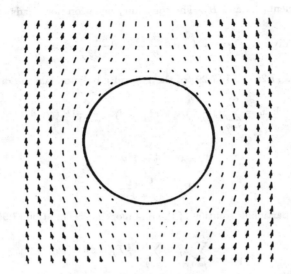

Fig. 7.1   Velocity field around the sphere

and
$$v(R, \theta) = 0 \tag{7.107}$$

yields

$$B_0 = -\frac{3}{2} v_0 R, \tag{7.108}$$

$$B_j = 0, \quad j = 1, 2, 3, ..., \tag{7.109}$$

$$C_0 = 0, \tag{7.110}$$

$$C_1 = -\frac{3}{2} \eta v_0 R, \tag{7.111}$$

$$C_j = 0, \quad j = 2, 3, 4, ..., \tag{7.112}$$

$$E_0 = 0, \tag{7.113}$$

$$E_1 = -\frac{1}{4} v_0 R^3, \tag{7.114}$$

$$E_j = 0, \quad j = 2, 3, 4, ... \tag{7.115}$$

All together

$$p = -\frac{3}{2} \eta v_0 \frac{R}{r^2} \cos \theta, \tag{7.116}$$

$$v_r = (1 - \frac{3}{2} \frac{R}{r} + \frac{1}{2} \frac{R^3}{r^3}) v_0 \cos \theta, \tag{7.117}$$

$$v_\theta = -(1 - \frac{3}{4} \frac{R}{r} + \frac{1}{4} \frac{R^3}{r^3}) v_0 \sin \theta. \tag{7.118}$$

The infinitesimal force dF^j acting on an infinitesimal surface element $d\Sigma_k$ in the fluid is given by the momentum flux π^{jk} through that surface:

$$dF^j = \sum_{k=1}^{3} \pi^{jk} d\Sigma_k. \tag{7.119}$$

The momentum flux is given by

$$\pi^{jk} = \rho v^j v^k + p\delta_{jk} - \sigma^{jk} \tag{7.120}$$

with

$$\sigma^{jk} := \eta \left(\frac{\partial v^j}{\partial x^k} + \frac{\partial v^k}{\partial x^j} \right) \tag{7.121}$$

in Cartesian coordinates or in polar coordinates

$$\sigma_{rr} = 2\eta \frac{\partial v_r}{\partial r}, \tag{7.122}$$

$$\sigma_{\theta r} = \eta \left(\frac{1}{r} \frac{\partial v_r}{\partial \theta} + \frac{\partial v_\theta}{\partial r} - \frac{v_\theta}{r} \right), \tag{7.123}$$

$$\sigma_{\phi r} = \eta \left(\frac{\partial v_\phi}{\partial r} + \frac{1}{r \sin \theta} \frac{\partial v_r}{\partial \phi} - \frac{v_\phi}{r} \right), \tag{7.124}$$

On the sphere we have

$$\sigma_{rr} = \sigma_{\phi r} = 0, \tag{7.125}$$

$$\sigma_{\theta r} = -\frac{3}{2} \eta \frac{v_0}{R} \sin \theta \tag{7.126}$$

and

$$\begin{aligned} dF^3(R, \theta) &= \frac{3}{2} \eta \frac{v_0}{R} \cos^2 \theta |d\Sigma| + \frac{3}{2} \eta \frac{v_0}{R} \sin^2 \theta |d\Sigma| \\ &= \frac{3}{2} \eta \frac{v_0}{R} |d\Sigma|. \end{aligned} \tag{7.127}$$

Finally integrating over the sphere we arrive at Stokes' law

$$F^3 = \frac{3}{2} \eta \frac{v_0}{R} 4\pi R^2 = 6\pi R\eta v_0. \tag{7.128}$$

Exercises

1. Show that the Laplace operator commutes with the gradient and with the divergence.
2. Prove equation (7.96).

3. Compute the coefficients (7.105) and (7.106) from the boundary condition (7.104).

4. Computethe coefficients (7.108-115) from the boundary condition (7.107).

5. Calculate the infinitesimal force (7.127).

7.5 Bessel functions

For $\ell = 0, 1, 2, \ldots$ consider the self-adjoint operator

$$A_\ell := -\frac{d^2}{dx^2} + \frac{\ell^2 - 1/4}{x^2} \tag{7.129}$$

with properly chosen domain of definition in $\mathcal{L}^2([0, \infty))$. Its spectrum, the positive half axis, is purely continuous. The function

$$e_\lambda(x) := \sqrt{\frac{\pi}{2}x}\ J_\ell(\sqrt{\lambda}x) \tag{7.130}$$

is eigendistribution with positive eigenvalue λ, where $J_\ell(x)$ is the Bessel function of order ℓ. It can be defined as ℓ-th Fourier coefficient of the periodic function in t

$$e^{-ix\sin t} =: \sum_{\ell=-\infty}^{+\infty} J_\ell(x)e^{-i\ell t}. \tag{7.131}$$

The first three Bessel functions are shown in fig. 7.2.

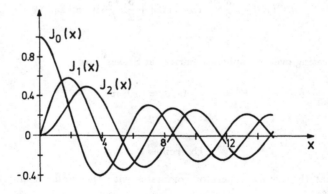

Fig. 7.2   Three Bessel functions

All Bessel functions are real (for real x). They have oscillatory character and possess infinitely many roots. The positive roots $a_{\ell m}$ of the Bessel function of index ℓ are numbered in increasing order by $m = 1, 2, 3, ...$ They are listed in tables of Bessel functions, e.g.

$$a_{01} = 2.4048, \tag{7.132}$$

$$a_{02} = 5.5201, \tag{7.133}$$

$$a_{11} = 3.8317, \tag{7.134}$$

$$a_{12} = 7.0156. \tag{7.135}$$

For large x the Bessel functions behave like trigonometric functions and the zeros become equi-distant:

$$J_\ell(x) \sim \sqrt{\frac{2}{\pi x}} \cos(x - \frac{\pi}{2}\ell - \frac{\pi}{4}). \tag{7.136}$$

Just as the plane waves for the operator $\mathrm{i}d/dx$, the eigendistributions (7.130) are not square integrable and they define a generalized Fourier transform, the so-called Hankel transform of index ℓ:

$$\hat{f}(k) := \frac{1}{\sqrt{\pi}} \int_0^\infty f(x) \sqrt{\frac{\pi}{2}x} \; J_\ell(\sqrt{k}x)dx \tag{7.137}$$

with inverse

$$f(k) := \frac{1}{\sqrt{\pi}} \int_0^\infty \hat{f}(k) \sqrt{\frac{\pi}{2}x} \; J_\ell(\sqrt{k}x)dk \tag{7.138}$$

where f is any square integrable function, $f \in \mathcal{L}^2([0, \infty))$.

Let us collect a few useful properties of the Bessel functions:

$$\left(x^2 \frac{d^2}{dx^2} + x\frac{d}{dx} + x^2 \right) J_\ell(x) = \ell^2 J_\ell(x) \tag{7.139}$$

called Bessel equation,

$$J_0(x) = \frac{1}{\pi} \int_0^\pi e^{\mathrm{i}x \cos t}dt, \tag{7.140}$$

$$J_\ell(x) = x^\ell \left(-\frac{1}{x}\frac{d}{dx} \right)^\ell J_0(x), \quad \ell = 1, 2, 3, ..., \tag{7.141}$$

$$J_{\ell+1}(x) = \frac{2\ell}{x} J_\ell(x) - J_{\ell-1}(x). \tag{7.142}$$

The operator on the lhs of Bessel's equation (7.139) is Hermitian on the space $\mathcal{C}_0^\infty((0, \infty))$ with respect to the scalar product

$$(f, g) := \int_0^\infty \bar{f}(x)g(x)\frac{dx}{x} \tag{7.143}$$

154

and it is tempting to believe that the Bessel functions form an orthogonal basis of $\mathcal{L}^2_{1/x}([0,\infty))$. This is not true:

$$\int_0^\infty J_\ell(x) J_{\ell'}(x) \frac{dx}{x} = \begin{cases} \frac{1}{2\ell}, & \ell = \ell' > 0 \\ \frac{\sin[(\ell-\ell')\pi/2]}{(\ell^2-\ell'^2)\pi/2}, & \ell \neq \ell'. \end{cases} \tag{7.144}$$

Furthermore, the Bessel functions J_ℓ can be defined for arbitrary real positive order ℓ using the self-adjoint operator (7.129) and all formulas in this section concerning J_ℓ remain valid in the general case except of course equation (7.141). The Bessel functions of half integer order for instance can be expressed in terms of trigonometric functions:

$$J_{1/2}(x) = \sqrt{\frac{2}{\pi}} \frac{\sin x}{\sqrt{x}}, \tag{7.145}$$

$$J_{3/2}(x) = \sqrt{\frac{2}{\pi}} \left(\frac{\sin x}{x\sqrt{x}} - \frac{\cos x}{\sqrt{x}} \right) \tag{7.146}$$

and so forth with the recurrence relation (7.142).

Exercises

1. Derive Bessel's equation from equations (7.129) and (7.130).
2. Prove equations (7.140-1).
3. Prove the recurrence relation (7.142) for integer $\ell = 1, 2, 3, \ldots$
4. Show that the only weight function $\rho(x)$ rendering the differential operator in Bessel's equation Hermitian on $C_0^\infty((0,\infty))$ is $\rho(x) = x^{-1/2}$.
5. Show that the Bessel functions of half integer order defined by equations (7.145-6) and the recurrence relations (7.142) satisfy Bessel's equation.

7.6 The circular membrane

In chapter 3 we have used the Fourier series to solve the Cauchy problem of the (finite) string. This procedure generalizes immediately to the rectangular membrane by separating the two spatial variables and yields plane waves. The dynamics of the circular membrane is more interesting. Consider the 2+1 dimensional wave equation in polar coordinates $r \leq 1$ and ϕ (2.16-17) with $c = 1$

$$\left(\frac{\partial^2}{\partial t^2} - \frac{\partial^2}{\partial r^2} - \frac{1}{r} \frac{\partial}{\partial r} - \frac{1}{r^2} \frac{\partial^2}{\partial \phi^2} \right) u(t, r, \phi) = 0 \tag{7.147}$$

with initial condition

$$u(0, r, \phi) = g(r, \phi), \tag{7.148}$$

$$\frac{\partial u}{\partial t}(0, r, \phi) = h(r, \phi) \tag{7.149}$$

and subject to the boundary condition (of Dirichlet type)

$$u(t, 1, \phi) = 0. \tag{7.150}$$

With the separation of variables

$$u(t, r, \phi) = v(t)w(\phi)q(r) \tag{7.151}$$

we obtain

$$v''/v = -\lambda, \tag{7.152}$$

$$w''/w = -K \tag{7.153}$$

with two real constants λ and K. The function $w(\phi)$ must be periodic with period 2π. Therefore K must be the square of an integer ℓ,

$$K = \ell^2, \quad \ell \geq 0. \tag{7.154}$$

Finally for $q(r)$ our ansatz yields Bessel's equation

$$r^2 q'' + r q' + \lambda r^2 q = \ell^2 q, \tag{7.155}$$

and

$$u(t, r, \phi) = J_\ell(a_{\ell m} r)(A \cos \ell\phi + B \sin \ell\phi)(G \cos a_{\ell m} t + D \sin a_{\ell m} t) \tag{7.156}$$

solves the wave equation with the boundary condition (7.150) where A, B, C and D are arbitrary real constants. As for the string, we now have to ask: Was the ansatz general enough so that a superposition of the above solutions satisfies the initial conditions and how to determine the constants? Again the answer is supplied by expanding the initial values into a (generalized) Fourier series. With respect to the periodic variable ϕ we take the ordinary real Fourier series, of course. With respect to r, $0 \leq r \leq 1$, the so-called Fourier-Bessel series of order ℓ solves the problem:

$$f(r) = \sum_{m=1}^{\infty} c_m J_\ell(a_{\ell m} r), \tag{7.157}$$

where f is a differentiable function that vanishes at $r = 1$. The coefficients c_m follow from scalar products:

$$\int_0^1 J_\ell(a_{\ell m} r) J_\ell(a_{\ell m'} r) r \, dr = \frac{1}{2}(J_{\ell+1}(a_{\ell m}))^2 \delta_{mm'}, \tag{7.158}$$

$$c_m = \frac{2}{(J_{\ell+1}(a_{\ell m}))^2} \int_0^1 f(r) J_\ell(a_{\ell m} r) r \, dr. \tag{7.159}$$

The main difference between plane waves $\cos kx$ and "cylindrical" waves $J_\ell(kr)$ is that only the former have equi-distant nodes. This is why the Fourier-Bessel series looks more complicated than the Fourier series. Also note that the possible frequencies of a circular drum

$$\nu_{\ell m} = \frac{a_{\ell m}}{2\pi} \frac{c}{R},$$

(7.160)

R the radius, c the speed of sound, are not integer multiples of the lowest frequency. The drum does not produce octaves to its keynote.

Exercises

1. Solve the Cauchy problem of the rectangular membrane. Are its possible frequencies equi-distant?

2. Determine the coefficients A, B, C and D as a function of ℓ and m in a superposition of solutions (7.156) such that the initial conditions (7.148) and (7.149) are satisfied.

3. Show that the functions

$$\sqrt{r} J_\ell(a_{\ell m} r), \quad r \in [0, 1]$$

are eigenfunctions of the differential operator A_ℓ (7.129). What are the eigenvalues? Now prove the orthogonality relation, i.e. (7.158) with $m \neq m'$.

References

Bonnor, W. B. (1969). Comm. Math. Phys. **13**, 163

Choquet-Bruhat, Y.,& DeWitt-Morette, C. (1989).
Analysis, Manifolds, and Physics. Amsterdam: North Holland

Dirac, P. A. M. (1930). The Principle of Quantum Mechanics. Oxford:
Oxford University Press

Feynman, R. P. (1948). Phys. Rev. **74**, 939

Glass, P., Hwang, D. H., & Sirlin, J. (1989).
1000 Air Planes on the Roof. Salt Lake City: Gibbs-Smith

Heaviside, O. (1893). Proc. Roy. Soc. **A52**, 504

Heaviside, O. (1894). Proc. Roy. Soc. **A54**, 105

Kirchhoff, G. (1982). Sitzungsberichte der Preußischen Akademie
der Wissenschaften, 641

Lighthill, M. J. (1958). Introduction to Fourier Analysis and Generalized Functions.
Cambridge: Cambridge University Press

Schwartz, L. (1950). Théorie des distributions. Paris: Hermann

Schwartz, L. (1965). Méthodes mathématiques pour les sciences physiques.
Paris: Hermann

Stückelberg, E. C. G. (1942). Helv. Phys. Acta **15**, 23

Talman, J. D. (1968). Special functions. A Group Theoretic Approach
New York: Benjamin

Temple, G. (1953). J. Lond. Math. Soc. **28**, 134

Temple, G. (1955). Proc. Roy. Soc. **A228**, 175

Bibliography

Abramowitz, M.,& Stegun, I. A. (1972). Handbook of Mathematical Functions
with Formulas, Graphs, and Mathematical Tables. New York: Wiley

Akhiezer, N. I.,& Glazman, I. M. (1961, 1963).Theory of Linear Operators
in Hilbert Space. New York: Frederick Ungar.

Bjorken, J. D.,& Drell, S. D. (1964). Relativistic Quantum Mechanics.
New York: McGraw-Hill

Bohm, A., & Gadella, M. (1989). Dirac Kets, Gamov Vectors, and Gelfand
Triplets: The Rigged Hilbert Space Formulation of Quantum Mechanics.
Berlin: Springer

Cartan, H. (1961). Théorie élémentaire des functions analytiques d'une ou
plusieurs variables complexes. Paris: Hermann

Constantinescu, F. (1974). Distributionen und ihre Anwendungen in der
Physik. Stuttgart: Teubner

Courant, R., & Hilbert, D. (1968). Methoden der Mathematischen Physik.
Berlin: Springer

Gelfand, I. M. & Shilov, G. E. (1964). Generalized Functions.
New York: Academic Press

Großmann, S. (1988). Funktionalanalysis. Wiesbaden: Aula

Guichardet, A. (1969). Calcul intégral. Paris: Armand Colin

Jones, D. S. (1966). Generalised Functions. London: McGraw-Hill

Lebedev, N. (1965). Special Functions and Their Applications.
Englewood Cliffs, N. J.: Prentice Hall

Magnus, W., Oberhettinger, F., & Soni, R. P. (1966). Formulas and Theorems
for the Special Functions of Mathematical Physics. New York: Springer

Richtmyer, R. D. (1978). Principles of Advanced Mathematical Physics.
New York: Springer

Scharf, G. (1989). Finite Quantum Electrodynamics. Berlin: Springer

Watson, G. N. (1966). A Treatise on the Theory of Bessel Functions.
Cambridge: Cambridge University Press

Zemanian, A. H. (1965). Distribution Theory and Transform Analysis.
New York: McGraw Hill

Notation

$\mathbb{N}$	the natural numbers
$\mathbb{Z}$	the integer numbers
$\mathbb{R}$	the real numbers
$\mathbb{C}$	the complex numbers
$\bar{z}$	complex conjugate of $z \in \mathbb{C}$, *3*
A^T	transpose of the matrix A
$A^\dagger$	Hermitian conjugate of the matrix A
	or adjoint of the operator A, *135*
$\mathbb{1}$	unit matrix or identity operator, *119*
$\otimes$	tensor product, *50*
$*$	convolution product, *46*
$\dot{}$	d/dt
$'$	d/dx
(a, b)	open interval $\{x \in \mathbb{R} \mid a < x < b\}$
$[a, b]$	closed interval $\{x \in \mathbb{R} \mid a \leq x \leq b\}$
$\lim_{x \to a+}$	limit as x approaches a from above
$\lim_{x \to a-}$	limit as x approaches a from below
$\propto$	proportional
$\delta(x)$	delta distribution, *32, 50*
$H(x)$	Heaviside's step function, *27*
$G_\xi(x)$	Green function, *43*
$T(fg)$	time-ordered product, *48*
r, θ, ϕ	polar coordinates in $\mathbb{R}^3$, *51*
r	radius in $\mathbb{R}^N$, *49*
c_j	complex Fourier coefficients, *59*
a_j, b_j	real Fourier coefficients, *55*
$e_j(t)$	complex harmonics, *61*

T	period, *56*
ν	frequency, *70*
ω	angular frequency, *70*
k	wave number or wave vector, *76, 92*
q	norm of wave vector, *93*
$\hat{f}(k)$	Fourier transform of the function $f(x)$, *78, 91*
S_N	area of unit sphere in $\mathbb{R}^N$, *98*
$x \cdot y$	Euclidean scalar product in $\mathbb{R}^N$, *91*
(x, y)	scalar product in a complex vector space, *121*
$|x|$	norm in $\mathbb{R}^N, \mathbb{C}^N$ or in a complex vector space, *122*
δ_{jk}	Kronecker symbol, *61*
$\eta_{\mu\nu}$	Minkowski metric, *92*
$P(k)$	polynomial of linear differential operator with constant coefficients, *96*
Δ	Laplace operator, *97*
$\Box$	wave operator, *104*
$\partial\!\!\!/$	Dirac operator, *114*
$\partial/\partial N_\xi$	normal component of the gradient with respect to ξ
u, v	light-cone coordinates, *100*
$\mathcal{C}^\infty(\Omega)$	vector space of smooth functions over Ω, *117*
$\mathcal{C}_0^\infty(\Omega)$	vector space of smooth distributions over Ω, *117*
$\mathcal{D}(\Omega)$	vector space of distributions over Ω, *117*
$\mathcal{S}(\mathbb{R}^N)$	vector space of functions of fast decrease, *119*
$\mathcal{T}(\mathbb{R}^N)$	vector space of tempered distributions, *121*
ℓ^2	Hilbert space of square summable sequences of complex numbers, *122*
$\mathcal{L}^2(\Omega)$	Hilbert space of square integrable "functions" over Ω, *133*
$\mathcal{L}_{\rho^2}^2(\Omega)$	Hilbert space of square integrable "functions" with respect to measure $\rho^2(x)dx$, *141*

$[A, B]$	commutator of two operators,	*119*
$\langle A \rangle$	expectation of Hermititan operator,	*125*
ΔA	standard deviation of Hermitian operator,	*125*
$\mathcal{D}(A)$	domain of definition of Hilbert-space operator,	*134*
$\Gamma(x)$	gamma function,	*93*
$h_j(x)$	Hermite functions,	*140*
$H_j(x)$	Hermite polynomials,	*141*
$L_j(x)$	Laguerre polynomials,	*141*
$P_j(x)$	Legendre polynomials,	*142*
$P_j^m(x)$	associated Legendre functions,	*144*
$T_j(x)$	Tschebysheff polynomials of first kind,	*143*
$U_j(x)$	Tschebysheff polynomials of second kind,	*143*
$Y_\ell^m(\theta, \phi)$	spherical harmonics,	*145*
$J_\ell(x)$	Bessel functions,	*152*

Index